T0142866

Models and Modeling in Science Education

Volume 11

More information about this series at http://www.springer.com/series/6931

Kristy L. Daniel
Editor

Towards a Framework for Representational Competence in Science Education

 Springer

Editor
Kristy L. Daniel
Texas State University
San Marcos, TX, USA

ISSN 1871-2983   ISSN 2213-2260   (electronic)
Models and Modeling in Science Education
ISBN 978-3-030-07900-0   ISBN 978-3-319-89945-9   (eBook)
https://doi.org/10.1007/978-3-319-89945-9

Printed on acid-free paper

This Springer imprint is published by the registered company Springer International Publishing AG part
of Springer Nature.
The registered company address is: Gewerbestrasse 11, 6330 Cham, Switzerland

Contents

Part I
The Importance of Representational Competence

Towards a Definition of Representational Competence

Kristy L. Daniel, Carrie Jo Bucklin, E. Austin Leone, and Jenn Idema

Currently, there is not a consensus in science education regarding representational competence as a unified theoretical framework. There are multiple theories of representational competence in the literature that use differing perspectives on what competence means and entails. Furthermore, dependent largely on the discipline, language discrepancies cause a potential barrier for merging ideas and pushing forward in this area. In science, representations are used to display data, organize complex information, and promote a shared understanding of scientific phenomena. As such, for the purposes of this text, we define representational competence as a way of describing how a person uses a variety of perceptions of reality to make sense of and communicate understandings. While a single unified theory may not be a realistic goal, strides need to be taken towards working as a unified research community to better investigate and interpret representational competence. Thus, this chapter will define aspects of representational competence, modes of representations, and the role of a representational competence theoretical framework in science education research and practice.

Science is often communicated via visual means, be it graphs, tables, models, diagrams, or simulations. This style of communication relies on the intended receiver's ability to make sense of the visual inputs in manners consistent with scientific thinking. With growing access to technology and its continued integration into

K. L. Daniel (✉) · J. Idema
Texas State University, San Marcos, TX, USA
e-mail: kristydaniel@txstate.edu; jli17@txstate.edu

C. J. Bucklin
Southern Utah University, Cedar City, UT, USA
e-mail: Carriebucklin@suu.edu

E. Austin Leone
Oklahoma State University, Stillwater, OK, USA
e-mail: ealeone@okstate.edu

© Springer International Publishing AG, part of Springer Nature 2018
K. L. Daniel (ed.), *Towards a Framework for Representational Competence in Science Education*, Models and Modeling in Science Education 11,
https://doi.org/10.1007/978-3-319-89945-9_1

educational environments (laptops, tablets, smartphones, desktop monitors, etc.), there is a growing trend that students are being exposed to even more science visualizations as a means of communicating ideas. It is becoming ever more critical to understand how to best help students learn using visualizations in science. Through investigating and discussing the role of representational competence in students' science learning we can better understand how to help students use science visualizations.

This book serves to initiate thinking about a representational competence theoretical framework across science educators, learning scientists, practitioners and scientists as well as provide a current state of thinking about representational competence in science education. Authors in the following chapters pose new questions to consider and explore ideas linked to representational competence in science education with regards to external visualizations as we press forward in advancing our field through research, and bridge thoughts across disciplines.

Background and Theory

Representations are useful tools that organize complex information, display data, and elaborate on complex topics in ways that make the information easier to understand. Representations are critical for communicating abstract science concepts (Gilbert 2005), where there can be vast amounts of data or phenomena that when written in text, can lead to misconceptions. Ignoring how students use and develop scientific representations will prevent them from developing expertise in their field. Rather we need to focus on understanding how students learn to interact and communicate with scientific representations.

There are two primary types of representations: external and internal. External representations are visually perceivable models while internal representations result from perceptions that remain inside the mind. The distinction between these representation types is sometimes blurred by the assumption that the focus of cognitive research is ultimately on internal representations. The classification of pictorial and verbal representations constructed by students can be used as an assessment tool to help represent internal representations or mental models held and used by students. All the same, understanding the nature and role of external representations in content areas is important when investigating instruction and learning, because external representations themselves can be a significant component of reasoning within that domain. When learning with representations, there are two frameworks that capture and explain STEM learning with visualizations as learners develop expertise, representational competence and model competence.

Representational competence is a way of describing how a person uses a variety of perceptions of reality to make sense of and communicate understanding through external visualization (Halverson and Friedrichsen 2013; Kozma and Russell 2005). To determine students' representational competence, representational fluency must also be addressed. Representational fluency is a measure of representational competence, and is the process of translating and moving within and between representations to understand a concept. While representational competence is static, representational fluency is the students' ability to seamlessly move within and between representations, ultimately increasing learning (see Chap. 2 for more detail).

Representations can be expressed though five external forms: concrete, verbal, symbolic, visual, and gestural (Gilbert 2005). These forms of fluency are explored more in depth in Chap. 2, in which the understanding of STEM concepts rests with the learner's ability to represent these concepts and then translate between and within representational forms. When a learner achieves high representational competence, he/she can begin shifting the external representation into an internal representation, or a mental image that can be manipulated to improve performance on visual tasks, memory tasks, and cognitive problem solving (Gilbert 2005; Botzer and Reiner 2005; Clement et al. 2005). Competence can then be investigated as an outcome, condition, or developmental stage with students' understanding of content based on their interactions with representations.

Where representational competence is a way of describing a learner's ability to use representations, model competence describes how a person interacts with a representation either as a medium (to interpret or illustrate meaning) or a method (a process to test or make predictions about questions) (Upmeier zu Belzen and Krüger 2010). This competence can be measured by capturing data on how effectively a person completes varying tasks (See Chap. 11 for further details). Developing high representational and model competence does not happen overnight. Students enter the classroom with preconceptions about many topics and these preconceptions can influence their understanding of how to internalize, interpret, and interact with external visualizations (Meir et al. 2007). It is possible for a learner's level of representational competence to change based on content (Gilbert 2005) and task difficulty. Therefore, it is critical to consider how students use and make sense of representations depicting science content.

Visual representations play a key role in mathematics, geography, and science (Cuoco and Curcio 2001; Gilbert 2005) and can be considered a means to form conceptual understanding (Zazkis and Liljedahl 2004). We know that visualizations can enhance learning from texts, improve problem solving, and facilitate connections between new knowledge and prior knowledge (Cook 2006). Various forms of visual representations can support an understanding of different, yet overlapping, aspects of a phenomenon or entity. While there is no doubt that the use of visual representations enhances learning (Cook 2006; Meyer 2001; Peterson 1994; Reiner and Gilbert 2008; Woleck 2001) students' ability to comprehend and interact with visual representations is often found lagging (Zbiek et al. 2007; Anderson and Leinhardt 2002; Ferk et al. 2003; Reiss and Tunnicliffe 2001; Tufte 2001). In science, the National Research Council (1996) outlined the following objectives for students working with representations. Students should be able to:

- Describe and represent relationships with visual representations;
- Analyze relationships and explain how a change in an entity affects another;

- Systematically collect, organize, and describe data;
- Describe and compare phenomena;
- Construct, read, and interpret representations;
- Support hypotheses and argument with data;
- Evaluate arguments based on data presented;
- Represent situations with multiple external visual representations and explore the interrelationship of these representations; and
- Analyze representations to identify properties and relationships.

When visual representations are understood accurately, they can provide depictions of phenomena that cannot be illustrated through other approaches. In some cases, visual representations can be misleading and create additional difficulties with interpretation (Zbiek et al. 2007; Tufte 2001). This is often the case when students use representations as a literal depiction of the phenomenon (Anderson and Leinhardt 2002). For example, children learning anatomy sometimes view symbolic references of a heart as a realistic expectation to how an anatomical heart will appear (Reiss and Tunnicliffe 2001). However, before we can fully begin to understand how students use and interact with representations we need to recognize the different ways different scientific content areas approach representations. The chapters in this book cover aspects of representational competence within multiple science domains.

A Chance to Reach Consensus

Understanding how representations are interpreted in different content areas begs the question, can we reach a consensus of how representations influence learning?

Implications for Thinking

Research has shown that students have difficulties identifying key structures of visual representations, interpreting and using visual representations, transitioning among different modes of representations (e.g., 2D and 3D models), relating abstract representations to content knowledge, and understanding what approaches are appropriate for making sense of representations (Ferk et al. 2003; Gabel 1999; Hinton and Nakhleh 1999; Johnstone 1993; Treagust et al. 2003). Sometimes, the way students make sense of a visualization may lead to correct responses, but this does not mean that the students have used an appropriate approach (Tabachneck et al. 1994; Trouche 2005). Experts are able to organize knowledge from visual representations into patterns that inform actions and strategies, while novice students rely upon superficial knowledge of equations and representations rather than patterns to generate solutions (Bransford et al. 1999; Larkin et al. 1980). For students to become experts with visualizations, they must learn how to interpret visualizations correctly and use them as a reasoning tool when investigating problems (Cavallo 1996).

An example of this difficulty in developing expertise with visualizations has been documented in evolutionary biology. In this domain, phylogenetic trees use branches and nodes to represent hypothesized relationships among species by mapping descent from common ancestry. Phylogenetic trees can be verbally explained to assist with interpretation. However, it is difficult to represent relationships among organisms at the same level of detail without using a visualization. For example, patterns of monophyletic groups (a common ancestor and all descended lineages, also referred to as a clade) and genetic algorithms (evolutionary computations used to identify optimality) are difficult to comprehend without a visual or symbolic image. This process of using a phylogenetic approach to understand evolutionary biology is referred to as tree-thinking (Baum and Smith 2013). Within tree-thinking, there are two core skill sets required for understanding these trees: tree reading and tree building (Halverson 2011).

Experts in systematics are identified by their ability to comprehend phylogenetic trees as representations of species relatedness and are able to use trees as reasoning tools when solving systematics problems. They use phylogenetic representations to interpret and illustrate patterns among the evolutionary histories of different species lineages. Thus, understanding phylogenetic trees involves overcoming prior naïve ideas about species and interpreting relations based on the branching patterns of the tree. It is imperative that students are able to interpret and recognize patterns when processing evolutionary trees. If students cannot recognize patterns within phylogenetic trees, then they will not be able to accurately interpret the intended meaning nor test the hypothesis presented. Thus, it is critical that learners are assisted with learning how to recognize patterns, particularly in scientific visualizations (Anderson and Leinhardt 2002; Tabachneck et al. 1994; Simon et al. 1989).

Interpreting visual phylogenetics representations often depends more on pattern recognition than on conceptual understanding. Given student explanations of their reasoning processes, they seem to misinterpret trees because of flawed reasoning in pattern recognition (e.g., associating species proximity to each other as relatedness) (Baum et al. 2005; Gendron 2000; Gregory 2008; Meisel 2010; Meir et al. 2007) or incorporating foundational misconceptions about evolution into tree thinking (Halverson et al. 2011; Gibson and Hoefnagels 2015; Walter et al. 2013). Part of the issue might be due to students not accessing information visually presented by a tree to make sense of a given problem. However, we are just now accessing biometric data that offers evidence for how experts or novices visually access information gleaned from tree diagrams.

Relatively new, emerging technology now allows investigators to gather biometric data about eye movement patterns and interactions with visual stimuli. The process of recording and measuring eye movement patterns, called eye-tracking, is used in a variety of disciplines (Duchowski 2002) and can aid representational competence researchers identify how information is visual accessed. Eye-tracking is frequently used in reading comprehension studies (Rayner 2009), but its application in science education is growing. For example, using eye-tracking technology to gather biometric data has helped evolutionary biologists understand how students are visually interacting with phylogenetic tree diagrams (Novick et al. 2012). Chapter 11 explores the use of eye-tracking as a means for assessing and understanding visual attention while using representations in science education. The information in Chap. 11 synthesizes prior use of this tool as well as providing insight towards future research for using eye-tracking as way to assess representational competence.

A more tradition way to assess students' level of representational competence is through qualitative methods. Interviewing students or collecting data via open-ended questionnaires can provide an in depth understanding of a student's level of representational competence. For example, qualitative methods were instrumental in documenting the inventory of common tree-thinking misconceptions (Gregory 2008; Halverson et al. 2011). However, qualitative means are not always time efficient for classroom use, nor when assessing large groups of students. In these instances, quantitative means would be more efficient. Quantitative assessment assigns a numerical value, usually on a scale, to indicate a level of progress or competency. In the case of tree-thinking, one skill that can easily be quantified in relation to competency is visualization of rotation. Mental rotation has been linked to students' ability to succeed in topics stemming from spatial ability (Bodner and Guay 1997). There is evidence that ability to think visually and manipulate images is linked to problem solving in chemistry (Stieff and Raje 2010; Stieff 2007). Unfortunately, visual-spatial thinking is often overlooked by science educators (Mathewson 1999). Although we see evidence of the role of mental rotation in many areas of science and medicine, only preliminary studies have quantitatively measured students' visualization of rotation using a multiple choice quantitative instrument (Bodner and Guay 1997) and examined at how it impacts learning with visualizations in biology (Maroo and Halverson 2011). Tree-thinking requires mental rotation skills as phylogenetic trees are often presented as two dimensional representations, but requires processing in three dimensional space in order to interpret and compare diagrams (Halverson 2010). This rotational aspect of these trees is comparable to that of molecular models seen in chemistry. And difficulties with mental rotation can lead to challenges in developing expertise in representational competence. More detail on assessing representational competence is discussed in Part III of this book.

Call to Measure Representational Competence across Disciplines

There is a need to investigate representational competence across domains. Information visualization is crucial for processes of high level cognition and communication (including education, decision-making, and scientific inquiry), with separate visualization traditions established in these respectful fields (Roundtree 2013; Skupin 2011). For example, in geography, visual depiction and analysis of spatial relationships is often the core of the scientific study in question, including natural and human phenomena. In biology, phylogenetic trees and maps of species distribution are meant to teach and communicate relationships between classes within kingdoms of species. In computer science, proximity of blocks of code and 3D animations can aid the learnability of interfaces, as well as illustrate interrelationships between computer program sub-functions. In chemistry, atomic models use proximity to depict visual representations of interactions between individual elements. Despite obvious parallels between these fields, no cross-domain theory of visual cognition and communication exists to date. Exploring ways to synthesize and evaluate a set of fundamental cross-discipline visualization principles that enable human comprehension and valuation, moving past research into visual perception towards the science of visual cognition would greatly improve visual communication design and education in natural, theoretical and applied sciences (Fabrikant and Skupin 2005; Rayl 2015).

Looking forward, there is also justification for research into ways visual communication can impact English Language Learners (ELL) and visually impaired students learning. Understanding how visualizations affect learning in these communities of students can provide a more thorough and inclusive learning environment in the classroom. Specifically, exploring how underrepresented student groups like ELL and visually impaired students interact with visualizations might provide educators with key insight for increasing learning opportunities for these students. For example, consider such use as 3D printing to create and adapt a manipulative model that would serve as a tactile aid for students who cannot see visual models or that may have other cognitive learning abilities. Such modifications could lead to increased representational competence for these students. Regardless of specific content area, there are many opportunities to investigate how representations are used in the classroom and how educators can integrate more inclusive representational forms for all students.

References

Anderson, K. C., & Leinhardt, G. (2002). Maps as representations: Expert novice comparison of projection understanding. *Cognition and Instruction, 20*, 283–321.

Baum, D. A., & Smith, S. D. (2013). *Tree thinking: An introduction to phylogenetic biology.* Greenwood Village: Roberts.

Baum, D. A., Smith, S. D., & Donovan, S. S. S. (2005). The tree-thinking challenge. *Science, 310*, 979–980.

Bodner, G. M., & Guay, R. B. (1997). The Purdue visualizations of rotations test. *The Chemical Educator, 2*, 1–17.

Botzer, G., & Reiner, M. (2005). Imagery in physics learning-from physicists practice to naive Students Understanding. In *Visualization in science education* (pp. 147–168). Netherlands: Springer.

Bransford, J. D., Brown, A. L., & Cocking, R. R. (1999). *How people learn: Brain, mind experience, and school*. Washington, D.C.: National Academy Press.

Cavallo, A. (1996). Meaningful learning, reasoning ability, and students understanding and problem solving of topics in genetics. *Journal of Research in Science Teaching, 33*(6), 625–656.

Clement, J., Zietsman, A., & Monaghan, J. (2005). Imagery in science learning in students and experts. In *Visualization in science education* (pp. 169–184). Netherlands: Springer.

Cook, M. P. (2006). Visual representations in science education: The influence of prior knowledge and cognitive load theory on instructional design principles. *Science Education, 90*, 1073–1091.

Cuoco, A. A., & Curcio, F. R. (Eds.). (2001). The roles of representation in school mathematics. National Council of teachers.

Duchowski, A. T. (2002). A breadth-first survey of eye-tracking applications. *Behavior Research Methods, Instruments, & Computers, 34*(4), 455–470.

Fabrikant, S. I., & Skupin, A. (2005). Cognitively plausible information visualization. In J. Dykes, A. M. MacEachren, & M.-J. Kraak (Eds.), *Exploring Geovisualization*. Amsterdam: Elsevier.

Ferk, V., Vrtacnik, M., Blejec, A., & Gril, A. (2003). Students understanding of molecular structure representations. *International Journal of Science Education, 25*, 1227–1245.

Gabel, D. (1999). Improving teaching and learning through chemistry education research: A look to the future. *Journal of Chemistry Education, 76*(4), 548.

Gendron, R. P. (2000). The classification & evolution of caminalcules. *American Biology Teacher, 62*, 570–576.

Gibson, J. P., & Hoefnagels, M. H. (2015). Correlations between tree thinking and acceptance of evolution in introductory biology students. *Evolution: Education and Outreach, 8*, 15.

Gilbert, J. K. (2005). *Visualizations in science education* (Vol. Vol.1). Dordrecht: Springer.

Gregory, T. R. (2008). Understanding evolutionary trees. *Evolution: Education and Outreach, 1*, 121–137.

Halverson, K.L. (2010). Using pipe cleaners to bring the tree of life to life. *American Biology Teacher, 74*, 223–224. (Associated Lesson Plan: http://dl.dropbox.com/u/4304176/ConferencePapers/PipeCleanerLessonPlan.doc).

Halverson, K. L. (2011). Improving tree-thinking one learnable skill at a time.*Education and Outreach Evolution*: Austin, *4*(1), 95–106.

Halverson, K. L., & Friedrichsen, P. (2013). Learning tree thinking: Developing a newFramework of Representational Competence. In D. F. Treagust & C.-Y. Tsui (Eds.), *Models and Modeling in Science Education, Multiple Representations in Biological Education* (Vol. 7, pp. 185–201). Dordrecht: Springer.

Halverson, K. L., Pires, C. J., & Abell, S. K. (2011). Exploring the complexity of tree thinking expertise in an undergraduate systematics course. *Science Education, 95*(5), 794–823.

Hinton, M. E., & Nakhleh, M. B. (1999). Students microscopic, macroscopic, and symbolic representations of chemical reactions. *The Chemical Educator, 4*(5), 158–167.

Johnstone, A. H. (1993). The development of chemistry teaching: A changing response to changing demand. *Journal of Chemistry Education, 70*(9), 701.

Kozma, R. B., & Russell, J. (2005). Modelling students becoming chemists: Developing representational competence. In J. K. Gilbert (Ed.), *Visualization in science education* (pp. 121–145). Dordrecht, The Netherlands: Springer.

Larkin, J., McDermott, J., Simon, D. P., & Simon, H. A. (1980). Expert and novice performance in solving physics problems. *Science, 208*, 1335–1342.

Maroo, J., & Halverson, K. L. (2011). Tree-Thinking: A branch of mental rotation. *Synergy: Different Entities Cooperating for a Final Outcome, 2*(2), 53–59.

Mathewson, J. H. (1999). Visual-spatial thinking: An aspect of science overlooked by educators. *Science Education, 83*, 33–54.

Meir, E., Perry, J., Herron, J. C., & Kingsolver, J. (2007). College students' misconceptions about evolutionary trees. *American Biology Teacher, 69*, 71–76.

Meisel, R. P. (2010). Teaching tree-thinking to undergraduate biology students. *Evolution: Education and Outreach, 3*(4), 621–628.

Meyer, M. R. (2001). Representation in realistic mathematics education. In A. A. Cuoco (Ed.), *The roles of representation in school mathematics (2001 Yearbook)* (pp. 238–250). Reston, VA: National Council of Teachers in Mathematics.

National Research Council. (1996). *National science education standards*. Washington D.C.: National Academy Press.

Novick, L. R., Stull, A. T., & Catley, K. M. (2012). Reading Phylogenetic Trees: The Effects of Tree Orientation and Text Processing on Comprehension. *BioScience, 62(8)*, 757–764.

Peterson, M. P. (1994). Cognitive issues in cartographic visualization. In A. M. MacEachren & D. R. F. Taylor (Eds.), *Visualization in Modern Cartography* (pp. 27–43). Oxford: Pergamon.

Rayl, R. (2015). Implications of Desnoyers' taxonomy for standardization of data visualization: A study of students' choice and knowledge. *Technical Communication, 62(3)*, 193–208.

Rayner, K. (2009). Eye movements and attention in reading, scene perception, and visual search. *The quarterly journal of experimental psychology, 62(8)*, 1457–1506.

Reiner, M., & Gilbert, J. K. (2008). When an image turns into knowledge: The role of visualization in thought experimentation. In J. K. Gilbert, M. Reiner, & M. Nakhleh (Eds.), *Visualization: Theory and practice in science education*. Dordrecht, The Netherlands: Springer.

Reiss, M. J., & Tunnicliffe, S. D. (2001). Students understandings of their internal structure as revealed by drawings. In H. Behrendt, H. Dahncke, R. Duit, W. Graber, M. Komorek, A. Kross, & P. Reiska (Eds.), *Research in science education - Past, present, and future* (pp. 101–106). Dordrecht, The Netherlands: Kluwer Academic Publishers.

Roundtree, A. K. (2013). *Computer simulation, rhetoric, and the scientific imagination: How virtual evidence shapes science in the making and in the news*. Lanham, MD: Lexington Books.

Simon, H. A., Larkin, J. H., McDermott, J., & Simon, D. P. (1989). Expert and novice performance in solving physics problems. In H. A. Simon (Ed.), *Models of thought* (Vol. 2, pp. 243–256). New Haven, CT: Yale University Press.

Skupin, A. (2011). Mapping texta. *Glimpse Journal, 7*, 69–77.

Stieff, M. (2007). Mental rotation and diagrammatic reasoning in science. *Learning and Instruction, 17*, 219–234.

Stieff, M., & Raje, S. (2010). Expert algorithmic and imagistic problem solving strategies in advanced chemistry. *Spatial Cognition & Computations, 10*, 53–81.

Tabachneck, H. J. M., Leonardo, A. M., & Simon, H. A. (1994). How does an expert use a graph? A model of visual and verbal inferencing in economics. In *Proceedings of the 16th annual conference of the Cognitive Science Society* (Vol. 842, p. 847).

Treagust, D., Chittleborough, G., & Mamiala, T. (2003). The role of submicroscopic and symbolic representations in chemical explanations. *International Journal of Science Education, 25(11)*, 1353–1368.

Trouche, L. (2005). An instrumental approach to mathematics learning in symbolic calculator environments. In *The didactical challenge of symbolic calculators* (pp. 137–162). US: Springer.

Tufte, E. R. (2001). *The visual display of quantitative information* (2nd ed.). Cheshire, CT: Graphics Press.

Upmeier zu Belzen, A., & Krüger, D. (2010). Model competence in biology class. *Journal of Teaching Methods on Natural Sciences, 16*, 41–57.

Walter, E. M., Halverson, K. L., & Boyce, C. J. (2013). Investigating the relationship between college students acceptance of evolution and tree thinking understanding. *Evolution: Education and Outreach, 6*, 26.

Woleck, K. R. (2001). Listen to their pictures: An investigation of children's mathematical drawings. In *The roles of representation in school mathematics* (pp. 215–227). Reston: National Council of Teachers of Mathematics.

Zazkis, R., & Liljedahl, P. (2004). Understanding primes: The role of representations. *Journal. for Research in Mathematics Education, 35*, 164–186.

Zbiek, R. M., Heid, M. K., Blume, G. W., & Dick, T. P. (2007). Research on technology in mathematics education: A perspective of constructs. In F. K. Lester (Ed.), *Second handbook of research on mathematics teaching and learning* (Vol. 2, pp. 1169–1207). Charlotte, NC: Information Age Publishing.

Representational Fluency: A Means for Students to Develop STEM Literacy

Tamara J. Moore, S. Selcen Guzey, Gillian H. Roehrig, and Richard A. Lesh

STEM and STEM Integration

The STEM disciplines are distinct, yet are also inherently connected. For the purpose of distinction, let's take an academic view of the disciplines. Each discipline asks different kinds of questions. The processes used to answer questions differ significantly. Scientists are interested in explaining phenomena that occur in the natural world; mathematicians work towards extending the man-made, abstract world of logic, structure, and patterns; engineers address human problems and needs through advancing the designed world. While science, mathematics, and engineering are the disciplines of STEM, technology is not an academic discipline. Yet, it is important to discuss the role of technology in STEM. There are two common definitions of technology. In the more inclusive view, technology is defined as the process by which humans modify their surroundings to fit their needs and desires (International Technology and Engineering Education Association (ITEEA) 2000). In order to meet needs, engineering, mathematics, and science play an integral role in the modification. The second definition focuses on artifacts of technology: computers, medicine, wind turbines, etc. The National Academy of Engineering [NAE] (2009) states that technology

> "includes all of the infrastructure necessary for the design, manufacture, operation, and repair of technological artifacts ... The knowledge and processes used to create and to

T. J. Moore (✉) · S. Selcen Guzey
Purdue University, West Lafayette, IN, USA
e-mail: tamara@purdue.edu; sguzey@purdue.edu

G. H. Roehrig
University of Minnesota, Minneapolis, MN, USA
e-mail: roehr013@umn.edu

R. A. Lesh
Indiana University, Bloomington, IN, USA

© Springer International Publishing AG, part of Springer Nature 2018
K. L. Daniel (ed.), *Towards a Framework for Representational Competence in Science Education*, Models and Modeling in Science Education 11,
https://doi.org/10.1007/978-3-319-89945-9_2

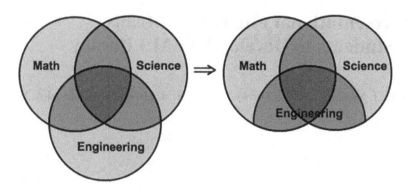

Fig. 1 Models of the STEM disciplines and how their content overlaps

operate technological artifacts – engineering know-how, manufacturing expertise, and various technical skills – are equally important part of technology."

Real-world problems often employ technologies, and creation or modification of technologies is the solution to the problems. For the purposes of this paper, we will view technology as an end product of the integration of the disciplines of STEM, and as such, the paper will focus heavily on science, mathematics, and engineering.

In Fig. 1, the left diagram represents the disciplines of mathematics, science, and engineering and their potential to overlap. When thinking about these three disciplines, it is easy to see this left diagram and agree that this could model the disciplines and their overlap. (Note, we are not making an inference to the size of these disciplines, just that overlap occurs.) Upon further reflection, however, one begins to question the existence of the discipline of engineering devoid of mathematics and science. Engineering without mathematics and/or science can be equated to tinkering. While sometimes an engineer might "tinker" on occasion, tinkering does not represent the discipline of engineering. Through close inspection of engineering, it can be said that the engineering discipline does not exist without mathematics and science. Therefore, we have adjusted the right diagram in Fig. 1 to represent this new perception. So once we consider the teaching of engineering, it is necessary to begin to think about integration of the disciplines. Even further reflection begins to question the overlaps of engineering/mathematics and engineering/science. Does engineering exist in these spaces? The engineering/mathematics space definitely does exist. Industrial engineers often use engineering to solve problems mathematically without using science principles or concepts. For example, consider the routes, departure times, etc. of planes from an airline company. The scheduling of all planes is an optimization problem that falls in the domain of industrial engineering. The underlying concepts for this engineering problem are mathematical, but not scientific. The existence of engineering/science space is not as clear. While an example has not presented itself to prove the existence of this space, we are not yet ready to dismiss its existence. However, it is clear that in reality, most of the discipline of engineering lies in the darkest gray space, the overlap of all three disciplines. The model of the disciplines of mathematics, science, and engineering shown in Fig. 1 is, of course, incomplete and too simplistic to represent the disciplines. There are many other factors that make up each discipline. For example, when addressing the disciplines of science and engineering, one must consider the role of ethics in their practice. However, for the purposes of this paper, we chose to ignore the other factors and concentrate just on the content and intersections of the STEM disciplines.

The model in Fig. 1, however, begs the question, "If engineering does not exist without mathematics and science, what makes it a separate discipline?" Engineering practice, at its core, is a way of thinking in order to solve problems for a purpose. This is characterized by the engineering design process, which is the "distinguishing mark of the engineering profession" (Dym 1999). According to ABET, the accrediting board for post-secondary engineering programs in the United States, engineering design is "a decision-making process (often iterative), in which the basic sciences, mathematics, and the engineering sciences are applied to convert resources optimally to meet these stated needs" (ABET 2017). Dym et al. (2005) define engineering design as "a systematic, intelligent process in which designers generate, evaluate, and specify concepts for devices, systems, or processes whose form and function achieve clients' objectives or users' needs while satisfying a specified set of constraints" (p. 104). A simplistic model of the engineering design process represents a problem solving process that goes through the steps (iteration implied) *Ask, Imagine, Plan, Create,* and *Improve* (Museum of Science Boston 2009). A more sophisticated model of engineering design is detailed in the study by Atman et al. (2007) on how engineering design experts and engineering students compare in their design processes. This paper states that there are five basic themes of design: (1) *problem scoping and information gathering,* (2) *project realization,* (3) *considering alternative solutions,* (4) *distributing design activity over time* (or *total design time and transitions*), and (5) *revisiting solution quality*. These views of design show that, like scientific inquiry, engineering design is not a lock-step procedure, but rather is a process in which decisions about what step in the design process to take next are made based on what was learned during the previous step. But it is clear from all of these views that science and mathematics play an integral role in engineering. ABET (2017) states this succinctly by saying, "the engineering sciences have their roots in mathematics and basic sciences but carry knowledge further toward creative application" (p. 2). From the other perspective, it could also be argued that much of the useful mathematics and science also lies in the spaces where mathematics, science, and engineering intersect. And, when they intersect, they tend to change in fundamental ways. If one takes this premise to be true, it follows that education of students in these disciplines is not completely representative unless integration of the subjects is meaningful.

Teachers of science and mathematics have been facing the challenges of teaching subject content in ways that engage students in meaningful, real-world settings (NGSS Lead States 2013; National Governors Association Center for Best Practices,, and Council of Chief State School Officers 2010). However, in many cases, a disconnect exists between "school science/mathematics" and "real science/mathematics." Engineering is a vehicle to provide a real-world context for learning science and mathematics. Any search for engineering education in K-12 settings will reveal a multitude of engineering outreach programs and curricular innovations. Institutions of higher education and professional associations provide compelling rationales for incorporating engineering in the precollege curriculum, either as a course in its own right or woven into existing mathematics and science courses. Common arguments for K-12 engineering education include (Koszalka et al. 2007; Hirsch et al. 2007):

- Engineering provides a **real-world context** for learning mathematics and science
- Engineering design tasks provide a context for developing problem-solving skills
- Engineering design tasks are complex and as such promote the development of **communication skills and team work**
- Engineering provides a **fun** and hands-on setting that will **improve students' attitudes** toward STEM careers.

While these are very good arguments for the inclusion of engineering the K-12 curriculum, a more powerful argument comes from the realization that the problems of the world are changing, as discussed at the start of this paper. Integration of engineering into mathematics and science courses makes sense given both the nature of the twenty-first century problems and the need to provide more authentic, real-world meaning to engage students in STEM. In order to prepare students to address the problems of our society, it is necessary to provide students with opportunities to understand the problems through rich, engaging, and powerful experiences that integrate the disciplines of STEM.

In 2000, Massachusetts became the first state to include engineering in their K-12 curriculum frameworks (Massachusetts Department of Education 2009). Other states, including Minnesota, Oregon, and Texas (Minnesota Department of Education 2009; Oregon Department of Education 2009; Texas Education Agency 2009), quickly followed. It is important to note that Minnesota, Oregon, Texas, and Massachusetts integrated engineering into their science standards rather than creating stand-alone engineering standards. This was a recommendation by the National Academy of Engineering and the National Research Council (NRC) through their report, *Standards for Engineering Education?* (NAE & NRC 2010). While allowing for the inclusion of stand-alone engineering courses, the intent of these documents is to integrate science and engineering. Following this trend, NRC followed these policy changes and recommendations, to put forth *A Framework for K-12 Science Education* (NRC 2012) which highlighted the need to bring engineering practices to the level of scientific inquiry as a way to teach science. Based on this call, the Next Generation Science Standards (NGSS Lead States 2013) took the framework and developed a comprehensive set of standards that integrate engineering into science learning throughout K-12 (Moore et al. 2015). To date, 19 states plus the District of Columbia have adopted NGSS as their science standards,11 states heavily used the NGSS for their craft their science education standards, and 5 states have explicit engineering standards (update to Moore et al. 2015).

As indicated throughout this chapter, real-world problems draw on multiple disciplines; integration is a natural occurring phenomenon in problem solving processes. However, this is different than the way that science, mathematics, and engineering are typically taught in the K-12 and post-secondary settings. Traditionally, teachers and curriculum unpack complex problems for students and misrepresent the STEM disciplines as distinct chunks of knowledge that students apply to artificial uni-variate problems - classic examples of this are word problems used in mathematics and some sciences where the real-world is involved to test students' ability to apply a pre-learned equation. Another common teaching strategy is to focus on the actions or processes of the disciplines, in other words the *doing* of STEM, to answer the questions of their field (i.e., inquiry in science, design in engineering, problem solving or proof in mathematics).

For example, in science, some approaches ask students to follow a set of prescribed steps to walk through the "scientific method." In engineering, design processes have been taught through tinkering with a product until it is within acceptable ranges. Within mathematics, problem solving has been taught by teaching heuristics such as "draw a picture" or "work backwards." In virtually every area of learning or problem solving where researchers have investigated differences between effective and ineffective learners or problem solvers (e.g., between experts and novices), results have shown that the most effective people not only *do* things differently, but they also *see* (or *interpret*) things differently. The attempts to represent the disciplines by their actions with a focus on "doing", not only lose the richness of the contexts in which each discipline operates, but also sorely misrepresent each discipline in such a way that it has potential to harm students' understanding and interest. STEM integration provides a more authentic way to engage students in meaningful and interesting problems.

Directions for STEM Integration

Effective practices in STEM teaching involve complex problem solving, problem-based learning, and cooperative learning, in combination with significant hands-on opportunities and curriculum that identifies social or cultural connections between the student and scientific/mathematical content. But STEM teaching needs to go beyond where it is toward a focus on what understandings are needed to improve STEM learning in the twenty-first century. Due to the nature of the changing problems, these new foci must center on STEM integration. We are proposing three needs to guide research around STEM integration:

1. Rich and engaging learning experiences that foster deep content understanding in STEM disciplines and their intersections are needed for students.
2. Most teachers have not learned disciplinary content using STEM contexts, nor have they taught in this manner, and therefore new models of teaching must be developed if STEM integration is to lead to meaningful STEM learning; and

3. There is a need for curricula that integrate STEM contexts to teach disciplinary content in meaningful ways that go beyond the blending of traditional types of understandings.

With this end-in-view, we will present a research framework that crosses both the STEM discipline boundaries and the principles outlined above. The following outlines the kinds of research questions that need to be addressed in order to advance the preceding principles of STEM integration.

Student Learning in STEM Integration

Student learning must be at the forefront of research in STEM integration, and research should be focused on the following kinds of questions: (a) What does it mean for students to "understand" important concepts or abilities in mathematics, the sciences, engineering, or the intersections of these disciplines? (b) How do these "understandings" develop? (c) What kind of experiences facilitate development? (d) How can these understandings be documented and assessed – while focusing on a broader range of deeper or higher-order understandings than those emphasized in traditional textbooks and tests? (e) When students learn via problem solving in meaningful contexts, what new types of understandings and abilities is it possible to cultivate? Developing models of student learning are an essential first step in research on teaching and curriculum development (Clements 2007). In his *Curriculum Research Framework,* Clements (2007) calls for the development and revision of curricular modules to be done in accordance with models of children's thinking and learning within the specific content domain.

Teaching with STEM Integration

A changing focus on student learning centered on the development of problem-solving and modeling within complex, multi-disciplinary situations requires new approaches to teaching. In order to foster innovative teaching practices, teaching with STEM integration needs to be understood. As stated earlier, most teachers in STEM disciplines have neither learned nor taught the STEM subjects in an integrated manner. This calls for research into how the teacher learns to teach in more powerful ways. Questions arise around areas of knowledge (both content and pedagogical), beliefs, and practice. These and other questions will need to be addressed as STEM integration moves forward: (a) What kind of content knowledge do effective STEM integration teachers need to be successful? Pedagogical knowledge? Pedagogical content knowledge? (b) What are the underlying beliefs of teachers about STEM disciplines and how to best teach them that lead to effective teaching practice in STEM integration? (c) How can we foster these beliefs in other teachers?

(c) How do effective teachers help students unpack, formalize, and abstract the STEM concepts that they have focused on in the STEM integration activities and modules?

Curriculum Development for STEM Integration

If student learning is to be meaningful, curriculum that integrates STEM concepts and contexts must be research-based and demonstrate that student learning is fostered. In educational research today, we often use curriculum as the "treatment" in our studies. But in terms of looking at STEM integration, we must actually research the curriculum design itself. The development of curriculum should be a user-centered design process. It needs to take in the needs of the learner, the implementer, and the needs of the greater STEM community. In order to create curriculum with these users in mind, research should address the following kinds of questions: (a) What types of activities and modules engage all students? Students of color? Female students? Students of different cultural heritage? (b) What are the "big ideas" that need to be addressed with students to help them succeed in the twenty-first century? (c) What concepts are being highlighted when a STEM integration activity or module is implemented? (d) How does a curricular innovation help students learn the STEM concepts? (e) How does teacher implementation of and student learning during a STEM integration activity differ based on the traditional course in which the implementation occurred (i.e., chemistry, calculus, physics, etc.)?

Unquestionably over the past several decades research in science education has been centered on notions of inquiry. Teachers' classroom practices and curricula in the U.S. are measured against definitions of inquiry laid out in the national reform documents (e.g., American Association for the Advancement of Science [AAAS] 1993; NRC 1996, 2012). In an international review of the science education literature (Abd-El-Khalick et al. 2004), it is clear that the centrality of inquiry within science education is a global phenomenon. The emphasis on inquiry is grounded in notions of how students learn and in social constructivist learning theory, and most science educators adhere to the belief that properly constructed inquiry laboratories have the potential to enhance student learning (Hofstein and Lunetta 1982, 2004). This firm belief has guided most research in science education. Unfortunately, while grounded in constructivist learning theory, the majority of the empirical research has considered the products of students' learning (often through standardized testing) rather than the processes of student conceptual development (von Aufschnaiter et al. 2008). Studies comparing inquiry to traditional practices have been inconclusive (Burkam et al. 1997; Cohen and Hill 2000; Von Secker 2002). What is missing from the current research approach is an understanding of the process of concept development (diSessa 2002) and an understanding of students' knowledge beyond the symbolic representations most commonly emphasized in standardized assessments. Models of student learning and concept development need to be reflective of the new kinds of problems described earlier in this paper that draw on multiple

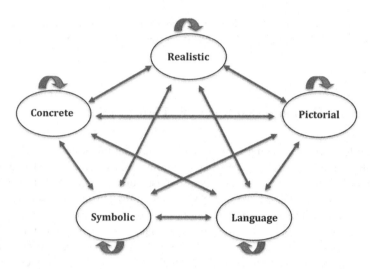

Fig. 2 The Lesh Translation Model (LTM)

STEM disciplines; in other words models of student learning need to be developed in the kinds of STEM integration settings that mirror real-world problems. While lacking a dynamic model of student learning, we cannot expect to explain how and why particular teaching strategies are successful at promoting student learning in STEM. We propose that representational fluency through the Lesh Translation Model, described in the following section, provides a framework for developing dynamic models of student learning and conceptual development in STEM integration that can be used as a model for research and teaching.

Lesh Translation Model as a Framework for STEM Integration

The Lesh Translation Model (LTM; Fig. 2) was originally designed to represent understanding of conceptual mathematical knowledge (Lesh and Doerr 2003). It consists of five nodes, each node is one category of representation: (1) Representation through Realistic, Real- World, or Experienced Contexts, (2) Symbolic Representation, (3) Language Representation, (4) Pictoral Representation, and (5) Representation with Manipulatives (concrete, hands-on models). The translation model emphasizes that the understanding of concepts lies in the ability of the learner to represent those concepts through the five different categories of representation, and that the learner can translate between and within representational forms (Cramer 2003).

Similar models to the LTM exist in science education. For example, three types of chemical representations are considered critical to a conceptual understanding of chemistry concepts: macroscopic (observable properties and processes), sub-microscopic or particulate (arrangement and motions of particles), and symbolic (chemical and mathematical notations and equations; Gabel 1998). A comparable model in physics education emphasizes the need to translate between a real-world situation, a free body diagram, and the mathematical models and symbols that describe the situation (Anzai 1991). There is overlap between these three models, yet the LTM is the most robust. So, we propose here that the LTM can be extended to science and engineering concepts and, therefore, provides an appropriate framework for research in STEM integration. In the following sections, we demonstrate the utility of this model in mapping student understanding, understanding of reform-based teaching, and curriculum development in STEM education.

The Lesh Translation Model and Student Learning

STEM integration allows new forms of understanding develop. Crossing STEM discipline boundaries cannot be simply the pouring together of each of the separate disciplines. As an example of weak integration of STEM, we can look to current classroom practices in science where mathematics is needed, but treated as a tool or an algorithm to be applied. Science draws heavily on mathematics, and traditional classroom practice and assessments have tended to focus heavily on this representation of science. Traditional wisdom has been that if the student can successfully complete math problems dealing with a particular concept in science, the student understands the underlying scientific concept or principle. But, several researchers have shown that students' ability to solve mathematical problems does not necessarily imply an understanding of the underlying concepts (Bodner 1991; Bunce and Gabel 2002; Nurrenbern and Pickering 1987). If we look to the LTM as a way to interpret understanding in STEM integration, we see that the model emphasizes the need for students to learn to translate between conceptual (language, pictorial) and symbolic representations, as well as the other representations. If students are to learn meaningfully, the representations of the concepts that students are able to produce should integrate the disciplines of STEM in a holistic manner. This mirrors "big picture" understanding rather than isolated pockets of knowledge that students must pull together like a patchwork quilt. In the next paragraphs, we will provide examples of weak integration that have to potential to be strong integration if the LTM is used as a framework to guide understanding.

A common concept in science that relies heavily on mathematics is the idea of pH. pH is a logarithmic function (pH = -log(concentration of hydrogen ions)) used to designate the level of acidity or alkalinity of a substance. Concepts involving logarithm functions are generally difficult for students to understand in the mathematics classroom. Mathematics teachers tend to present logarithms as an abstract concept without tying it to concrete, real-world applications.

These teachers tend to focus on the symbolic and graphic representations of this concept. This is further exacerbated by science teachers who do not teach the mathematics of logarithms during pH. When the need to have students deal with the logarithmic nature of pH arises, a common way to help students address this is to teach them which buttons on their scientific calculators are needed to produce the desired result without regard to what it conceptually means. These are missed opportunities for student learning in both the mathematics and science classrooms. If we want to take a STEM integration approach to pH and logarithms, students need multiple meaningful representations of the concepts of pH and the corresponding concepts of logarithms. The LTM can serve as the guide for the breadth of representations needed and the development of the translations between those representations.

Student learning opportunities are also missed in the area of scientific and engineering laboratory work. The LTM manipulative or hands-on representations, which include laboratory work, hold a sacred place in the science curriculum; indeed few science teachers and policy- makers would question the belief that students should experience a significant amount of laboratory work. Research in this area has tended to focus on barriers to implementing laboratory work (time, money, lab equipment, and space), debate on the degree to which laboratories should be student-centered (open vs. guided inquiry), and debate about the teaching of the nature of science (and more recently the nature of engineering) through the implicit laboratory-based experiences vs. explicit instruction. In reviewing the role of experiments in science instruction, Hodson (1988), argues that "the actual performance of the experiments contributes very little" to student learning. Unfortunately, this critique parallels the common translation by practitioners of inquiry as "hands-on" science where students engage in process activities devoid of content (Magnusson et al. 1999) or laboratory activities that focus on data collection with minimal or no attention to data analysis and the development of explanations from data, which is where significant STEM integration could occur. Too frequently, laboratory activities overlook the final three essential features of inquiry: the formulation of explanations from evidence, the connecting of explanations to that of the larger scientific knowledge, and the effective and justifiable communication of conclusions (NRC, 2000). Similarly, Brophy et al. (2008) caution that care must be taken that engineering activities do not end simply with the creation and evaluation of a product. If STEM learning goals are to be met, then students must evaluate alternative solutions and explain why they work. One could argue that the purpose of most laboratory activities implemented in K-12 classrooms serve the purpose of "edu-tainment" - they provide the "wow" and excitement of science without contributing directly to the learning of conceptual knowledge (Hodson 1988). For example, there is a buzz of excitement when the teacher holds the magnesium ribbon in the Bunsen flame or immerses a balloon in liquid nitrogen (traditional demonstration of Charles' Law). While engaging and memorable, these hands-on representations of science alone would not lead students to a conceptual understanding of chemical reactions. But careful attention to the representations beyond the manipulative node of the LTM and corresponding translations can potentially guide both the development of STEM explanations and arguments that connect to data (von Aufschnaiter et al. 2008) and STEM integration and, in turn, foster deep content understanding.

Allowing students to learn in situations that have them grapple with realistic problems that require crossing disciplines is the heart of STEM integration. Using engineering as a context for these problems is a natural way to allow students to learn through STEM integration. Engineering requires the use of scientific and mathematical concepts to address the types of ill-structured and open-ended problems that occur in the real world (Sheppard et al. 2009). Real-world engineering problems are complex with multiple viable solutions because of the number of variables and interrelationships between variables that need to be analyzed and modeled. Often, a client or end-user, either real or fictitious, needs to use the solution or design for a purpose. Therefore, the questions students investigate are guided by the client or user's needs and wants. While discussions about inquiry all focus on the need for students to be investigating "scientifically-oriented question(s)" leading to an understanding of natural phenomena (NRC 2000), debate about the authenticity of these questions has focused on who derives the question (student vs. teacher) and the types of questions that students tend to investigate are predicated by the assumption that students should first learn concepts and problem solving processes and only then should they put these two together to solve "real life" problems. But the questions that students choose to investigate are at best pseudo-real-world and univariate, and concepts develop through the process, not as separate entities that can be patched together. Engineering can provide a real-world context for STEM learning if and only if a STEM integration approach is taken. What must come to the forefront in classroom instruction is teachers' instructional practices that allow for and scaffold the development of explanations based on designs and data generated in the early stages of an inquiry or engineering design. Using engineering contexts as spaces for students to develop real- world representations can be the catalyst for developing related scientific and mathematical concepts through using LTM representations and facilitating translations. Teachers must facilitate discussions that engage students and value their ideas while moving students toward the development of appropriate understandings and explanations of STEM concepts.

The Lesh Translation Model and Teaching

Teachers, no matter what discipline they teach, need to have well-developed knowledge and deep understanding of their content area in order to teach it effectively. However, as Shulman (1986, 1987) suggests knowing content is not enough, teachers also need to know how to represent their content to students from different ages and backgrounds. STEM integration presents another issue related to teacher content knowledge in that their knowledge must expand beyond their discipline. Shulman points out that pedagogical content knowledge (PCK), which is a type of content knowledge, is a necessary knowledge base for teachers in order to present the content in more meaningful ways. A critical element of PCK is using different representations in teaching to help students in developing deep conceptual knowledge. Shulman suggests the use of representations including "the most powerful analogies, illustrations, examples, explanations, and demonstrations – in a word, the ways of representing and formulating the subject that makes it comprehensible for others" (1986, p. 9). According to Shulman, there is not only one type of most effective representation; thus, teachers need to have "at hand a veritable armamentarium of alternative forms of representation" and apply the most powerful ones to make the concepts more understandable to students (1986, p. 9). The LTM clearly indicates the different representations (manipulatives, pictures, realistic contexts, language, and symbols) that STEM teachers need to employ. However, as the LTM suggests "representational fluency" (the ability to fluently translate among and between representations) is a key element (Lesh 1999); having a large repertoire of representations is insufficient, teachers need to know how to scaffold students' ability to translate between representations. A mathematics teacher, for example, can represent concepts of ratio and proportion, as well as other "big ideas" in mathematics, through the use of gears in wind turbines. The context of redesigning wind turbines for the wind engineers that need to optimize power output based on the size of gear turned by the blades and size of gear that turns the generator allows students to develop understandings about proportion and ratio that are tied to a *real-world* context. A necessary piece of this is helping students understand the basics of electric generators, power, current, and voltage. This goes beyond what is typically considered a mathematics teacher's responsibility, so new understandings of teaching must develop. The teacher can also provide students with gear *manipulatives* and a working scale model of a wind turbine (with interchangeable gears) to allow them to begin to visualize the concepts and provide a place for hands-on experimentation with gear ratios. By providing the students with context and a problem that must be solved, the students can then investigate, experiment, inquire, and design which will promote development of other representations. An example of a problem that allows student to investigate gear ratio is to provide students with the prompt, "The engineers would like your team to help them be able to identify speed of any secondary gear based on any primary gear's speed. They need to be able to do this for any two gears, so please provide the engineers with enough information to understand how to apply your procedure to any two gear sizes." Teachers can then encourage student to represent their solutions through *pictorial representations* (e.g., diagrams), *symbolic representations* (e.g., equations), and *language representation* (e.g., a letter to the engineers explain how to go about using their procedure for any two gears). The LTM suggests that in order for students to deeply understand this or any other concept, students should both experience multiple representations but also learn to make connections between and among these representations (Cramer 2003; Lesh et al. 1987).

Once students have participated in a rich learning activity, it becomes the teacher's job to help students unpack what they have learned and help them abstract and decontextualize the concepts. A teacher's role in leading discussion becomes critical in implementing the LTM. The model suggests the teacher should "wrap up" or review all the representations in such a way that is comprehensible to students and allows them to develop the ability to translate between translations. Through these discussions, teachers can help students draw connections among multiple representations and build meaningful relationships among different classroom activities. Ball (1993) states that teachers to learn how to gauge whether important ideas are generated and be prepared to re-direct the conversation as needed. The role of the teacher in assisting students to develop arguments is critical as they need to guide students in making sense of the learning task as well as aligning students' ideas with accepted understandings in STEM. Unfortunately, teachers find this task difficult (Ball 1993; Leinhardt and Steele 2005; Schoenfeld 1998; Sherin 2002) and often inquiry lessons are left hanging as teachers accept students' ideas without evaluation (Furtak 2006). Stein et al. (2008) described practices that teachers should employ to enable generative discussions that included careful selection and sequencing of students' representations to be shared during discussion to build conceptual understanding, unpack the contextual ideas from activities, and help the class make connections between different students' responses and the key content ideas. The LTM can be used by teachers of different disciplines as a guide. The model has high potential in helping teachers in developing their PCK. Knowing about content representations makes a teacher different than a content specialist. As shown in various studies teachers with lack of knowledge on content representations cannot present the content properly (Borko and Putnam 1996; Magnusson et al. 1994; Zembal-Saul et al. 2000). Teachers can use the model in developing lesson plans or curricula. Using the five different modes of representations while focusing on the translation among them should be a focus of every teacher. In addition, teachers who try to integrate STEM disciplines into their curricula can use the LTM. Each discipline can be presented through applying multiple modes of representations facilitating translation not only between representations in a single content area but translations across all STEM disciplines.

The Lesh Translation Model and Curriculum Development

The use of standards-based curriculum materials has been widely promoted as a mechanism for teaching reform in STEM education, specifically when accompanied by comprehensive professional development (Powell and Anderson 2002; Post et al. 2008; Harwell et al. 2007). A new wave of standard-based curricular materials have been developed to cover topics in a more conceptual manner, tying together big ideas and themes in science or mathematics rather than focusing on the teaching of "factoids". There is evidence that teachers' use of standards-based NSF-funded mathematics and science curricula is related to higher student achievement (e.g., Burkam et al. 1997; Cohen and Hill 2000; Von Secker 2002), yet the evidence is not statistically significant. As indicated earlier in this paper, the gold standard of evaluation - running a traditional and reform-based curriculum in a "horse-race" with a standardized end-of-course test - will not resolve questions about the implementation and effectiveness of these curricula to promote student learning. We note here a need for new curricula that integrate the STEM disciplines, as well as a need for new directions in the frameworks that guide curriculum development and research into student learning with these new curricula. The LTM, which has been used successfully to guide curriculum development in mathematics (Cramer 2003), provides a framework for STEM integration curriculum development. It can be applied with curricular emphases being stronger in different representations to meet specific content learning goals, yet with all translations being present.

Critically, it is a model built on student learning and thus provides a bridging model or framework between student learning and curriculum development. The LTM also provides a framework for teachers to identify weaknesses in their existing curriculum and practices and a model to supplement absent representations as needed (Cramer 2003).

In STEM education, few realistic and complex problems are likely to be solved by single studies – or even by research based on single theoretical perspectives. So, it is highly significantly to take note of the fact that, in two recent handbooks surveying and analyzing research in mathematics education (Lester 2007) and science education (Abell and Lederman 2007), many leading STEM researchers identified "lack of accumulation" as the foremost shortcoming of math/science education research and development. And, "lack of communication" is one obvious cause of this lack of accumulation. That is, STEM education research consists of many subfields, which rarely interact, communicate, or build on one another's work. These subfields include not only the broad categories of mathematics education, chemistry education, physics education, biology education, and engineering education – but also even smaller sub-subfields such as those focusing on problem solving, or teacher development, curriculum materials development, or specific types of concept development (e.g., early number concepts, fractions and proportional reasoning, geometry and measurement, early algebra or calculus concepts, and so on). Consequently, even though there is evidence that STEM education researchers have made significant progress toward understanding what it means for students to "understand" some of the most powerful elementary-but-deep concepts in many of the preceding topic areas, the picture becomes far less clear when we notice that students' understandings of many of these concepts develop during overlapping periods of time. Yet, the theories which have evolved to explain these parallel developments often involve striking mismatches or incompatibilities. Similarly, Clements (2007) argues that the isolation of curriculum development and educational research are the reason that curriculum development has not reliably improved. In this paper, Clements (2007) argues for a Curriculum Research Framework that both uses and informs research on children's thinking and learning and research on teaching. The LTM provides a framework that interconnects models of STEM integration for student learning, teaching, and curriculum development and, due to the nature of the kinds of understandings it implies, suggests the utilization of a design-based research approach for STEM integration.

For learning to be transferable and applicable to the needs of the twenty-first century work environment, the kinds of problems that we use in classrooms and research need to mirror the nature of real-word problems; problems with which students engage need to integrate concepts and thinking across the STEM disciplines. The current practice of compartmentalizing problems into limited single discipline heuristics has lead to weak models of STEM integration that treat problem-solving as a pouring together of the disciplines rather than a true integration or reconceptualization of problem-solving that truly addresses real-word, multi-disciplinary problems. The LTM provides a framework for the structure of STEM that students need to have that demonstrates representational fluency between the multiple representations of STEM concepts. The LTM provides a consistent model for research in student thinking in STEM, teaching practices for STEM integration, and curriculum development in STEM integration.

References

Abd-El-Khalick, F., Boujaoude, S., Duschl, R., Lederman, N. G., Mamlok-Naaman, R., Hofstein, A., Niaz, M., Treagust, D., & Tuan, H. (2004). Inquiry in science education: International perspectives. *Journal of Research in Science Teaching, 32*, 397–419.

Abell, S. K., & Lederman, N. G. (Eds.). (2007). *Handbook of research on science teaching.* Mahwah, NJ: Lawrence Erlbaum Associates.

ABET. (2017). General criterion 5. Curriculum. Retrieved from http://www.abet.org/accreditation/accreditation-criteria/criteria-for-accrediting-engineering-programs-2017-2018/#outcomes

American Association for the Advancement of Science. (1993). *Benchmarks for science literacy.* New York: Oxford University Press.

Anzai, Y. (1991). Learning and use of representations for physics expertise. In K. A. Ericsson & J. Smith (Eds.), *Toward a general theory of expertise: Prospects and limits* (pp. 64–92). Cambridge, MA: Cambridge University Press.

Atman, C. J., Adams, R. S., Cardella, M. E., Turns, J., Mosborg, S., & Saleem, J. (2007). Engineering design processes: A comparison of students and expert practitioners. *Journal of Engineering Education, 96*(4), 359–379.

Ball, D. L. (1993). With an eye on the mathematical horizon: Dilemmas of teaching elementary school mathematics. *Elementary School Journal, 94*(4), 373–397.

Bodner, G. M. (1991). I have found you an argument: The conceptual knowledge of beginning chemistry graduate students. *Journal of Chemical Education, 68*, 385–388.

Borko, H., & Putnam, R. T. (1996). Learning to teach. In D. C. Berliner & R. C. Calfee (Eds.), *Handbook of education psychology* (pp. 673–708). New York: Macmillan.

Brophy, S., Klein, S., Portsmore, M., & Rogers, C. (2008). Advancing engineering education in P-12 classrooms. *Journal of Engineering Education, 97*, 369–387.

Bunce, D. M., & Gabel, D. (2002). Differential effects on the achievement of males and females of teaching the particulate nature of chemistry. *Journal of Research in Science Teaching, 39*, 911–927.

Burkam, D. T., Lee, V. E., & Smerdon, B. A. (1997). Gender and science learning early in high school: Subject matter and laboratory experiences. *American Educational Research Journal, 34*, 297–331.

Clements, D. H. (2007). Curriculum research: Toward a framework for "research-based curricula". *Journal for Research in Mathematics Education, 38*, 35–70.

Cohen, D., & Hill, H. (2000). Instructional policy and classroom performance: The mathematics reform in California. *Teachers College Record, 102*, 294–343.

Cramer, K. (2003). Using a translation model for curriculum development and classroom instruction. In R. Lesh & H. Doerr (Eds.), *Beyond constructivism: Models and modeling perspectives on mathematics problem solving, learning, and teaching* (pp. 449–463). Mahwah, NJ: Lawrence Erlbaum Associates.

diSessa, A. A. (2002). Why "conceptual ecology" is a good idea. In M. Limo'n & L. Mason (Eds.), *Reconsidering conceptual change: Issues in theory and practice* (pp. 29–60). Dordrecht: Kluwer.

Dym, C. (1999). Learning engineering: Design, languages and experiences. *Journal of Engineering Education, 88*(2), 145–148.

Dym, C., Agogino, A. M., Eris, O., Frey, D. D., & Leifer, L. J. (2005). Engineering design thinking, teaching and learning. *Journal of Engineering Education, 94*(1), 103–120.

Furtak, E. M. (2006). The problem with answers: An exploration of guided scientific inquiry teaching. *Science Education, 90,* 453–467.

Gabel, D. L. (1998). The complexity of chemistry and implications for teaching. In B. J. Fraser & K. G. Tobin (Eds.), *International handbook of science education* (pp. 233–248). Dordrecht: Kluwer Academic Publishers.

Harwell, M. R., Post, T. R., Maeda, Y., Davis, J. D., Cutler, A. L., Anderson, E., & Kahan, J. A. (2007). Standards-based mathematics curricula and secondary students performance on standardized achievement tests. *Journal for Research in Mathematics Education, 38*(1), 71–101.

Hirsch, L. S., Carpinelli, J. D., Kimmel, H., Rockland, R., & Bloom, J. (2007). The differential effects of pre-engineering curricula on middle school students' attitudes to and knowledge of engineering careers. Published in the proceeding of the 2007 Frontiers in Education Conference, Milwaukee, WI.

Hodson, D. (1988). Experiments in science and science teaching. *Educational Philosophy and Theory, 20,* 53–66.

Hofstein, A., & Lunetta, V. N. (1982). The role of the laboratory in science teaching: Neglected aspects of research. *Review of Educational Research, 52,* 201–217.

Hofstein, A., & Lunetta, V. N. (2004). The laboratory in science education: Foundations for the twenty-first century. *Science Education, 88*(1), 28–54.

International Technology Education Association. (2000). *Standards for technological literacy: Content for the study of technology.* Reston, VA: International Technology Association.

Koszalka, T., Wu, Y., & Davidson, B. (2007). Instructional design issues in a cross-institutional collaboration within a distributed engineering educational environment. In T. Bastiaens & S. Carliner (Eds.), *Proceedings of world conference on E-learning in corporate, government, healthcare, and higher education 2007* (pp. 1650–1657). Chesapeake, VA: AACE.

Leinhardt, G., & Steele, M. D. (2005). Seeing the complexity of standing to the side: Instructional dialogues. *Cognition and Instruction, 23*(1), 87–163.

Lesh, R. (1999). The development of representational abilities in middle school mathematics. In I. E. Sigel (Ed.), *Development of mental representation: Theories and applications* (pp. 323–350). Mahwah, NJ: Lawrence Erlbaum Associates.

Lesh, R., & Doerr, H. (2003). Foundations of a models and modeling perspective on mathematics teaching, learning and problem solving. In R. Lesh & H. Doerr (Eds.), *Beyond constructivisim: A models and modeling perspectives on mathematics problem solving, learning and teaching* (pp. 3–33). Mahwah, NJ: Lawrence Erlbaum Associates.

Lesh, R., Post, T., & Behr, M. (1987). Representations and translations among representations in mathematics learning and problem solving. In C. Janvier (Ed.), *Problems of representations in the teaching and learning of mathematics* (pp. 33–40). Hillsdale, NJ: Lawrence Erlbaum.

Lester, F. K., Jr. (Ed.). (2007). *Second handbook of research on mathematics teaching and learning.* Charlotte, NC: Information Age Publishing.

Magnusson, S., Borko, H., & Krajcik, J. (1994). Teaching complex subject matter in science: Insights from an analysis of pedagogical content knowledge. Paper presented at the 1994 Annual meeting of the National Association for Research in Science Teaching, Anaheim, CA..

Magnusson, S., Krajcik, J., & Borko, H. (1999). Nature, sources and development of pedagogical content knowledge for science teaching. In J. Gess-Newsome & N. G. Lederman (Eds.), *Examining pedagogical content knowledge* (pp. 95–132). Dordrecht: Kluwer Academic Publishers.

Massachusetts Department of Education. (2009). *Current curriculum frameworks: Science and technology/engineering.* Retrieved January 2, 2009 from http://www.doe.mass.edu/frameworks/current.html.

Minnesota Department of Education. (2009). *Academic standards in science: Draft two complete.* Retrieved January 2, 2009 from http://education.state.mn.us/MDE/Academic_Excellence/Academic_Standards/Science/index.html

Moore, T. J., Tank, K. M., Glancy, A. W., & Kersten, J. A. (2015). NGSS and the landscape of engineering in K-12 state science standards. *Journal of Research in Science Teaching, 52*(3), 296–318. https://doi.org/10.1002/tea.21199.

Museum of Science Boston. (2009). *Engineering is elementary engineering design process.* Retrieved April 15, 2009 from http://www.mos.org/eie/engineering_design.php.

National Academy of Sciences, National Academy of Engineering, and Institute of Medicine of the National Academies. (2006). *Rising above the gathering storm: Energizing and employing America for a brighter economic future.* Washington, DC: National Academies Press.

National Academy of Engineering. (2009). *What is technology?* Retrieved April 14, 2009 from http://www.nae.edu/nae/techlithome.nsf/weblinks /KGRG-55A3ER.

National Academy of Engineering, & National Research Council. (2010). *Standards for K-12 engineering education?* Washington, DC: National Academies Press.

National Center on Education and the Economy. (2007). *The report of the new commission on the skills of the American workforce.* San Francisco, CA: Jossey-Bass.

National Governors Association Center for Best Practices, & Council of Chief State School Officers. (2010). *Common core state standards for English language arts and mathematics.* Washington, DC: Author.

National Research Council. (1996). *National science education standards.* Washington DC: National Academy Press.

National Research Council. (2000). *Inquiry and the national science education standards: A guide for teaching and learning.* Washington DC: National Academy Press.

National Research Council. (2012). *A framework for K-12 science education: Practices, crosscutting concepts, and core ideas.* Washington, DC: The National Academies Press.

NGSS Lead States. (2013). *Next generation science standards: For states, By states.* Washington, DC: National Academic Press. Retrieved from http://www.nextgenscience.org/.

Nurrenbern, S. C., & Pickering, N. (1987). Concept learning versus problem solving: Is there a difference. *Journal of Chemical Education, 64*(6), 508–510.

Oregon Department of Education. (2009). *Oregon science K-HS content standards.* Retrieved April 27, 2009 from http://www.ode.state.or.us/teachlearn/subjects/science/curriculum/2009feb-adopted-k-h-science-standards.pdf.

Post, T. R., Harwell, M. R., Davis, J. D., Maeda, Y., Cutler, A., Anderson, E., Kahan, J. A., & Norman, K. W. (2008). Standards-based mathematics curricula and middle-grades students performance on standardized achievement tests. *Journal for Research in Mathematics Education, 39*(2), 184–212. .

Powell, J., & Anderson, R. D. (2002). Changing teachers' practice: Curriculum materials and science education reform in the USA. *Studies in Science Education, 37*, 107–135.

Schoenfeld, A. S. (1998). Toward a theory of teaching-in-context. *Issues in Education, 4*(1), 1–95.

Sherin, M. G. (2002). When teaching becomes learning. *Cognition and Instruction, 20*(2), 119–150.

Sheppard, S. D., Macantangay, K., Colby, A., & Sullivan, W. M. (2009). *Educating engineers: Designing for the future of the field.* San Francisco, CA: Jossey-Bass.

Shulman, L. S. (1986). *Those who understand: Knowledge growth in teaching* (pp. 4–14). February: Educational Researcher.

Shulman, L. S. (1987). Knowledge and teaching: Foundations of the new reform. *Harvard Educational Reviews, 57*, 1–22.

Stein, M. K., Engle, R. A., Smith, M. S., & Hughes, E. K. (2008). Orchestrating productive mathematical discussions: Five practices for helping teachers move beyond show and tell. *Mathematical Thinking and Learning, 10*, 313–340.

Texas Education Agency. (2009). *Curriculum: Texas Essential Knowledge and Skills: Science TEKS.* Retrieved January 2, 2009 from http://www.tea.state.tx.us/teks/scienceTEKS.html.

von Aufschnaiter, C., Erduran, S., Osborne, J., & Simon, S. (2008). Arguing to learn and learning to argue: Case studies of how students' argumentation relates to their scientific knowledge. *Journal of Research in Science Teaching, 45*, 101–131.

Von Secker, C. (2002). Effects of inquiry-based teacher practices on science excellence and equity. *Journal of Educational Research, 95*, 151–160.

Zembal-Saul, Z., Blumenfeld, P., & Krajcik, J. (2000). Influence of guided cycles of planning, teaching, and reflection on prospective elementary teachers science content representations. *Journal of Research in Science Teaching, 37*(4), 318–339.

Similar Information, Different Representations: Designing a Learning Environment for Promoting Transformational Competence

Billie Eilam and Shlomit Ofer

Transformation Among Representations and Modalities (TARM) is an important meta-representational competency (diSessa 2004) to foster learners' and users' management of complex information, especially data represented visually. However, like for other visualization abilities, school practices for promoting students' TARM abilities are scarce. The present chapter focuses on the unique design of a learning environment that aimed to promote TARM in fourth-grade girls. We describe this environment, which provided young dance students with long-term experiences (over a full academic year) in representing and transforming variably presented information on human body movements. We discuss the basic considerations and rationales for this environment's design, as well as the roles played by its distinct components in promoting the girls' TARM abilities. Finally, we present and discuss representative examples from the girls' transformational efforts over the year in light of the intended design. We demonstrate cases of successful environmental design that afforded the girls' advancement and cases of constraining design that impeded the girls' progress. We complete the chapter with a discussion of this environment's educational implications.

B. Eilam (✉)
University of Haifa, Haifa, Israel
e-mail: beilam@edu.haifa.ac.il

S. Ofer
The Kibbutzim College, Tel-Aviv, Israel

© Springer International Publishing AG, part of Springer Nature 2018
K. L. Daniel (ed.), *Towards a Framework for Representational Competence in Science Education*, Models and Modeling in Science Education 11,
https://doi.org/10.1007/978-3-319-89945-9_3

What Is Transformation?

Transformation is defined in different ways by *Merriam Webster's* online dictionary (Transform 2014): "to change in composition or structure," "to change the outward form or appearance of," "to change in character or condition," "to convert," "to change (something) into a different form so that it can be used in a different way," and "to change to a different system, method, etc." These definitions are adequate for our purpose in using this concept of transformation in the context of the present chapter, where we refer to transformation as including both the representation of a referent and also the translation of some representations and modalities into others. We use the acronym TARM to refer to Transformation Among Representations and Modalities.

Specifically, in the present study we refer to "transformation" as the operation whereby individuals transform or convert an information *source* – i.e., information that they perceive about a referent or information that is represented in a certain representational form or modality – into an information *target,* thereby representing that information in a different representational form or modality. In other words, this core transformational ability enables individuals to communicate information conveyed by a *source referent* or *representation* by means of a different *target representation.* The target itself may latter on serve as a source for a further refinement of the generated representation while creating a novel target. Examples for such transformations might be the conversion of: (a) a source data table on inter-city distances into a target map of the relevant geographical area with the distances marked; (b) an orally presented source description of a municipality's departmental structure into a target hierarchical flowchart diagram; (c) a source list of numerical botanical data on plant heights in different seasons into a target graph; or (d) video-taped source data of a chemical reaction between two verbally labeled liquids into a target abstract chemical symbolic formula. Within the context of the current research on young students' TARM abilities related to human body movement, examples might include converting an actual live or videotaped dynamic human motion into a static schematic representation, or transforming live-motor or graphic representation of a specific mode of human movement into a verbal-conceptual description (e.g., "hand forward").

The ability to transform representations has been discussed already using different terms. For example, diSessa et al. (1991) examined sixth-grade students' "translation" of represented motion among different types of graphs, position, velocity, and acceleration. They reported that "children improved and moved fluidly among a diverse collection of representational forms" (p. 147). In his 2004 study, diSessa also referred to the "modification" of representations as a core meta-representational competency. In our case, we use the term "transformation" to broadly include all possible transformations of observed source referents' or representations' forms or modalities into target representations.

Undeniably, valuable target representations may never be entirely identical to source referents or representations. If identicalness were desirable, transformation would be entirely redundant. Indeed, a change, reduction, or specification of information occurs during transformation, both when representing information about a referent and when converting between representational forms and modalities (Bertin 2007). Changes may also be induced due to divergent aspects of information being emphasized by different symbolic languages that may affect users' perceptions of the represented information, such as verbal texts, 3-D models that can be manipulated tactilely, or 2-D visual images. Moreover, different perceptual processes (e.g., attention) and modes of processing and storing information may be employed when those diverse modalities and symbolic languages are encountered.

Transformational Affordances

The important affordances of the TARM cognitive operation lie in the support it provides for achieving several different everyday aims and educational goals: enhancing accessibility, emphasizing/complementing peripheral/tacit/missing aspects of source information, organizing information, and limiting interpretations. First, TARM generally increases individuals' *accessibility* to certain information. For example, a tourist's filming of a live tribal dance can afford later audiovisual access to ephemeral information by converting it into long-lasting digital documentation. Likewise, juxtaposing a graphic diagram of a machine to a textbook, newspaper, or brochure can enhance readers' visual access to information by transforming the verbally described complex machine structure into a spatial illustration. Vice versa, a complex graphic diagram drawn on the classroom blackboard (i.e., visual spatial format) may be more easily understood if accompanied by the teacher's added oral explanation (i.e., auditory textual format). Similarly, TV newscasters may enhance viewers' ability to comprehend a complex atmospheric phenomenon exhibited in a photograph of outer space by presenting a schematic drawing, thereby transforming between different representational forms within the same modality.

Second, TARM may *emphasize/complement* peripheral/tacit/missing aspects of source information (i.e., source referent or representation) by highlighting/representing them in the target representation. The choice of target representational forms and modalities can thereby specify details, clarify ambiguities, complement information, or reduce "noise" (i.e., irrelevant information), which all are useful for promoting users' information processing and management. For example, the construction of a graph may afford the optimal method to convey trends or gaps over time when attempting to transform a great amount of data lists collected over many years (i.e., TARM within the same visual modality).

Third, TARM may result in a different *organization* of the information, thus exhibiting new/different referent's characteristics. For example, a hierarchical concept map, describing a specific topic, may provide a new perspective and convey new information on this topic when the same involved concepts are re-arranged in a different hierarchical order (i.e., TARM within the same visual modality). Likewise, transforming a complex written text into several tables that organize its details using different categories (rows) and their different values (columns) can allow for comparisons that were not afforded by the text.

Fourth, TARM may help to *limit interpretations* of different kinds of information conveyed by the source (Ainsworth 2006), which is particularly useful in cases where diverse interpretations may be made. For example, oral travel directions given to a tourist may be confusing, but transformation to a schematic map indicating specific landmarks may help limit the tourist's interpretation of the route by adding spatial information that is ambiguous when expressed verbally (i.e., TARM among different representational forms and modalities). An additional example is a 3D physical model of the biology cell system that may be confusing regarding the different layers comprising the system. A schematic drawing or a table representing the system levels each with its inherent components, may constraint the interpretation of this model structure and function (i.e., TARM between different representational forms in the same modality).

Transformational Constraints

Despite its many affordances, a successful performance of TARM may be constrained by students' knowledge deficits and by the learning environment's characteristics. The educational system typically lacks sufficient student experiences and teacher practices that promote effective representational knowledge and skills for transforming between different representational forms and modalities (Eilam 2012; Eilam and Gilbert 2014). Students must have knowledge about the different components and dimensions of a representation and their related considerations. They have to identify the represented referents, and be aware of the many representational forms (e.g., film, schematic illustration, map, diagram, graph) and their potential modalities (e.g., motoric, haptic, auditory, visual). Students also have to be aware of the distinct symbols (e.g., iconic, abstract, conventional, sounds) composing these representational forms and modalities and of their symbolic languages and grammars (e.g., text, image, movement, verbal). Finally, deficits in the domain knowledge relevant to the source referent or representation may result in students' failure to identify missing aspects of the source information or failure to differentiate between core and peripheral information.

Learners' ability to gain academically from TARM's advantages necessitates effortful development of different prerequisite skills that support TARM. For example, individuals must be able to compare the strengths and weaknesses of diverse possible target representations – namely, to map between them – in order to best represent the greatest amount of source information or its most important aspects as unambiguously as possible. Such ability would enable students' choice of the representational form and modality that are best suited to the intended purpose as well as their usage of this selected representation to improve accessibility and information management. Yet, these mapping abilities were found to challenge most students (Kozma 2003; Seufert 2003).

Specifically, TARM may be constrained by deficits in several basic skills, regarding the different dimensions involved in any representation like, the spatial and temporal dimensions. The spatial dimension use requires specific abilities (e.g., mental rotation, 2D-3D transformation, perspectives taking of birds-eye versus external versus internal observer's view [Tversky 2005]) and the temporal dimension use involves time-related abilities and considerations (e.g., related to static versus dynamic representations, time units and resolution, cross-sectional versus longitudinal data). Another possible constraining factor in developing TARM abilities is the inherent complexity of some source referents/representations, such as size scales that cannot be perceived by the human eye; complex structures and systems that may impose a high cognitive load; or an abstract or ephemeral source that is difficult to capture. Particularly tricky in the present case of promoting TARM for human movements are: the requisite conceptual domain knowledge about the human body and motor functions, the need for familiarity with spatial symbolic language and for well-developed mental rotation ability, and the ephemeral nature of the source information (fleeting body movements).

Last, representational forms/modalities may be presented on different mediums (e.g., paper, computer screen, TV, audio-recorder). In our case we have chosen to represent transformational activities using paper and pencil (Van Meter 2001) as well as motoric.

Therefore, similarly to other representational abilities, TARM capacities are not acquired through mere verbal abstract discussion but rather through repeated, varied practical experiences with diverse representations and rich contexts (diSessa and Sherin 2000). Individuals' many informal everyday experiences with representations may support their development of a rich foundation of intuitions and abilities related to TARM. However, the mindful application and further enhancement of TARM require intentional, formal, well-designed, and organized practices aiming to promote students' TARM-related knowledge base (e. g., Azevedo 2000; Bamberger 2007). For this purpose, TARM must be developed in an effective learning environment that provides these foundations by offering structured opportunities for appropriate experiences conducive to TARM knowledge construction.

Judging Representational Quality

In any research attempting to trace progress in transformational ability, judgments must be made concerning the quality of the produced target representations. According to diSessa (2002), "tradeoffs" always exist among multiple representational alternatives, with no perfect outcomes. Therefore, judgments of the target's representational quality must consider various factors including the task demands and the representation's stated goals. In the literature, several researchers have investigated young students' judgments of representational quality, albeit not necessarily in the context of developing transformational ability (e.g., diSessa 2004; Pluta et al. 2011). Interestingly, diSessa (2002) reported that his students' criteria for judging self-generated representations were similar to those of researchers in 90% of cases. He described "formal criteria" that considered systematicity, consistency, simplicity, redundancy, conventionality, and clear alignment of different parts of the representation. In an earlier study, diSessa et al. (1991) reported that sixth graders used formal criteria while self-generating representations of motion like, representational completeness regarding the represented referent information, compactness, precision regarding quantitative information, parsimony, economical use of unnecessary symbols, and learnability – easiness to explain. Verschaffel et al. (2010) confirmed diSessa's criteria in elementary school children who selected an appropriate representation for ephemeral sonic stimuli. Pluta et al. (2011) reported that seventh graders generated a large number of distinct epistemic criteria for judging the quality of scientific physical 3D models, on three levels: (a) primary criteria, focusing on the accuracy of the model with regard to its correspondence with the referent, as expected in science; (b) secondary criteria, contributing to epistemic scientific aims; and (c) ambiguous criteria suggesting students' misconceptions about the practices of science. Many seventh graders defined the models' goal as providing and explaining information, emphasized the models' clarity to achieve communicability, and underscored the models' parsimony – their appropriate amount of detail and complexity (Pluta et al. 2011).

In the current research, beyond such criteria based on the representations themselves and their goals, students' judgments of representations were highly influenced by our unique environmental design. Our research task required fourth-grade students to develop transformed representations of human movements that could be deciphered by peers. Thus, our young representation developers' main criteria for judging those representations during the transformation process focused on the young decipherer's expected ability to interpret the target representation by correctly performing the intended body movement. In this sense, they are similar to Pluta et al.'s (2011) primary criteria. This is not to say that other criteria like systematicity or precision were not considered in the course of the transformational process, as evidenced by some of the representation developers' comments recorded during small-group discourse. However, such explicitly expressed representation-oriented criteria aimed to solve particular local social or communicational problems that emerged during transformation, rather than to evaluate the product as a whole. Due to these environmental constraints, we adapted diSessa's (2004) term of *functional niche* – a context that makes certain demands on the representations produced in it – to describe how our students' main criteria for judging a representation were constrained by the task's functional demands. In sum, in this unique environment, students' criteria were mostly constrained to judgments about representations' communicativeness rather than focusing more broadly on "formal" representational quality.

We next describe our research methodology, including our participants, our designed learning environment for promoting TARM competence, and the rationales and considerations underlying its design. We continue by presenting our findings, focusing on several examples of students' transformational products and discussing how they were afforded or constrained by different aspects of the learning environment.

Method

Participants

We chose to examine TARM with fourth-grade girls ($N = 16$, ages 9 to10 years), whose motor development stage (Gallahue and Ozmun 1998), spatial ability stage (Hammill et al. 1993), and informal and formal meta-representational knowledge and experiences accumulated along their childhood (Sherin 2000) would enable them to perform the transformational tasks. An examination of the girls' school curricula and our acquaintance with Israeli culture supported our assumption that the girls had many previous opportunities, although mostly unintentional or inexplicit, for exposure to representations. Participation in the research was entirely voluntary after obtaining parental consent. Thirteen of the 16 girls completed their participation throughout the yearlong research, one dropped out and the remaining two did not participate consistently.

Overall Research Context

Over the course of one academic year, participating girls received a theoretical and experiential curriculum and were asked to perform various transformations using 11 increasingly complex "source" sequences of live and videotaped ephemeral human body movements demonstrated by the teacher. Each sequence transformations required 1 to 4 weekly video-recorded lessons, lasting 60 min each, held during afterschool hours in the school studio. In the first session, the 16 girls self-organized into four permanent groups, each sitting in a separate corner of the studio. For each of the 11 sequences, as seen in Table 1, the intervention included several phases:

(a) curricular instruction, including both verbal-conceptual and motoric instruction; (b) source sequence motor demonstration and self-generation of target visual-graphical representation (developers only); (c) decipherers' decoding and performance-based feedback to developers, and (d) developers' refinement of target representations, if necessary. Girls rotated roles between "developers" (three girls per group) and the "decipherer" (one girl per group).

Research Phases

Phase 1: Theoretical and experiential (verbal-conceptual and motoric) instruction Each novel movement sequence began with a multidimensional theoretical and experiential curriculum for all participants. Human movement involves the dimensions of body, time, and space, thereby greatly challenging acquisition of TARM. The topic of human movement as a "source" was selected due to its richness and complexity, which may provide diverse transformational experiences, as presented in Table 1. The curriculum's conceptual framework was based on an adapted version of the Eshkol-Wachman Movement Notation (Eshkol and Wachman 1958;

Table 1 Transformational Activities among Representational Forms and Modalities Experienced Within the Different Research Phases

Research phase	Information source	Information target	Representational form	Experienced transformation Modality	Example
1. Theoretical and experiential instruction	Verbal-conceptual	Physical-motor	Ephemeral-verbal→ ephemeral-motor	Auditory → motoric	Teacher says "raise leg forward" → girlsraise leg forward (motoric)
	Physical-motor	Verbal-conceptual	Ephemeral-motor → ephemeral-verbal	Motoric → verbal	Girls perform movement of raising leg forward → girls say "leg forward"
	Visual	Verbal-conceptual	Visual → ephemeral-verbal	Visual → verbal	Girls view live/ video demonstrating leg raised → girls say "leg forward"
2. Source sequence demonstration and self-generation of target representation[a]	Visual (observed live/video movement sequence)	Visual-graphic	Ephemeral-visual → enduring-visual-graphic	Motoric → visual-graphic	Developers view teacher's live/video-demonstrated leg raised forward → developers create visual-graphic symbolic drawing of leg moving forward

(continued)

Table 1 (continued)

				Experienced transformation	Example
3. Decipherment and performance-based feedback[b]	Visual-graphic	Motor (enacted live movement sequence)	Enduring-visual-graphic → ephemeral-motor	Visual-graphic → motor	Decipherer views peers' visual-graphic symbolic representation of leg raised forward → decipherer physically performs decoded movement (e.g., erroneously stepping forward, not raising)
4. Refinement of self-generated representation (optional)[a]	Ephemeral-motor and visual-graphic	Visual-graphic	Ephemeral-motor and previous enduring-visual-graphic → refined enduring-visual-graphic	Motoric and visual-graphic → visual-graphic	Developers compare their previous visual-graphic symbolic representation to decipherer's performance → developers refine symbolic representation to create clearer one of leg raised forward

Enduring-visual-graphic refers to the visual representation on paper. Phases 3 and 4 were repeated until developers judged the final target (decipherer's demonstrated movement sequence) as identical to the original source (teacher's demonstrated movement sequence)
[a]Developers only, with girls' roles rotated in each group
[b]Decipherers only, with roles rotated

Ofer 2009) and on the concept of "movement literacy" (Ofer 2001). Our curricular framework comprised four multi-element dimensions, which can help learners to meaningfully dismantle whole, complex, continuous movements into their component dimensions and discrete elements. These four dimensions can enable, for example, transformations of movements into sequences of static core segments in a preferred resolution, highlighting the dynamic changes occurring between every two consecutive segments (Tufte 1997).

The four multi-element dimensions were: (a) the **body-relative direction dimension**, referring to the body's movement as a whole unit and including the six spatial elements of forward, backward, right, left, up, and down; (b) the **absolute**

spatial direction dimension, also referring to the body's movement as a whole unit but with directions represented by eight elements (the numbers 0–7) in a compass rose-like form; (c) the **body parts dimension**, referring to independent movements of human parts (e.g., head, forearm, feet), according to their respective joint structure and aforementioned directions; and (d) the **temporal dimension**, referring to the sequencing of consecutive movement segments with different levels of complexity. In particular, three aspects of the temporal dimension were emphasized in instruction because they may specifically challenge participants: (i) representation of time as an abstract, ephemeral aspect of the dynamic movement and its transformation into a concrete, static graphic representation; (ii) transformation of the source's continuity into the target's distinct segments of a selected resolution; and (iii) identification of separate consecutive movements that occur one after the other in separate time units versus simultaneous several movements that occur concurrently within the same time unit.

In the current research, we applied these four multi-element dimensions to conceptualizing human movement, developing the curriculum, and analyzing the girls' transformational products. To enhance acquisition of these dimensions and their corresponding elements, they were presented in the curriculum with gradually increasing complexity, starting with a single dimension where the body moved as a whole unit and where each unit of movement equaled a unit of time, and eventually reaching multi-dimensional multi-element movements occurring simultaneously within and across time units. We aimed at the high-level conceptualization of each theoretical dimension's/element's (as a concept rather than as an example of it) independently from a specific experiential enactment (Ofer 2009). To enable it, the girls acquired the curriculum content through diverse verbal-conceptual and motoric training modes such as verbally defining theoretical dimensions/elements; translating verbal instructions into performed physical actions (e.g., "Step left"); and repeating training of each concept in various movement modes (e.g., jumping left vs. walking left vs. moving a foot left). This limited but accumulating body of instructed theoretical/experiential knowledge created a pool of shared information for learners to utilize while transforming representations, thereby constraining interpretational possibilities for developers, decipherers, and researchers. At the end of this phase, to verify learners' acquisition of the new verbal-conceptual and movement content for the particular sequence and to prevent knowledge deficits from intervening in participants' application of TARM, the teacher assessed the girls' knowledge of dimensions/elements after each learning unit, and gaps were narrowed if needed.

Phase 2: Source sequence demonstration and self-generation of target representation (developers only) For this phase, the four blind decipherers (one per group) exited the studio. Next, the four groups of developers viewed the teacher's silent live motor demonstration of the novel movement sequence, in both frontal and rear views to avoid developers' need for mental rotation. The difficulties involved in transforming an observed motoric information source into a visual-graphic target representation include the need to quickly construct a mental model of the movement before its disappearance. Our design promoted the girls' ability to overcome this difficulty by providing them with a video clip of the frontal and rear views of the source movement, which they could observe repeatedly at any time on a laptop computer. While observing the live/video demonstrations, some developers took notes. No verbal description or any form of graphical representation accompanied sequence presentation. Dynamic live/video representations were selected as the source information because they have the potential to support learners' construction of mental models for an ephemeral phenomenon like movement. They can explicitly demonstrate unnoticed details of movement as they unfold over time, thus supporting students' understanding. Moreover, the presentation of simultaneous coordinated changes in different moving human parts enhances students' ability to link between them to generate a single representation of movement (Zhang and Linn 2011). Similarly, ephemeral information that represents continuous chronological events of the same process enables a clear view of each event while its previous one disappears; this may increase attention and focus on the observed distinct event within the continuous process. Such transformations are not common practice in schools (see Bamberger 2007 for an example); therefore, we chose a design that provided such experiences.

After observing the source movement sequence's demonstration, the developers in each group were asked to collaboratively self-generate a target representation – a visual symbolic graphical representation on paper, with minimal verbal hints. The need for communicating the movement information to others was emphasized. This process was carried out in groups, in unsupported conditions (Van Meter 2001) with minimal teacher guidance. Rectangular A4 blank sheets of paper, pencils, and erasers were provided to the groups, whose discourse was videotaped. A self-generation task was selected to provide experiences in diverse transformational modes (here, and also in the refinement phase; see Table 1). Self-generation tasks require students to observe phenomena longer, more carefully, and more deeply while granting meaning to what they see (Zhang and Linn 2011). Moreover, self-generation was shown to engage students' conceptually more than ready-made given representations (Lehrer et al. 2000). In addition, the task of self-generation rather than giving participants existing conventional representations (e.g., ready-made graphs or diagrams) aimed to enable the girls to utilize their own resources (Sherin 2000) (e.g., drawing abilities, familiar symbols, the trained verbal-conceptual curriculum).

Phase 3: Decipherers' decoding and performance-based feedback to developers Decipherers next returned to the studio without seeing the original live/video ephemeral-motor source, and each decipherer received her group's symbolic graphical representation, which now constituted a source for the decipherer. While decipherers interpreted this representation, developers were not present, to exclude any hints or communication. To promote our analysis of each decipherer's considerations while interpreting her visual-graphic source, decipherers were asked to "think aloud" and were video-recorded during the decoding process. Decoding was based on decipherers' curriculum-acquired knowledge, their accumulating experience with developers' representations, their accumulating experiences as developers (due to rotating roles), and their informal and formal prior knowledge and experiences regarding symbolic languages. After decoding, the developers reentered the studio, and each decipherer, while thinking aloud, transformed her visual-graphic source representation back into a physically performed movement sequence (the target) for developers' inspection.

Phase 4 (Optional): Developers' target representational refinement phase As developers observed the decipherer's live motor feedback, they compared their perceptions of the decipherer's performance with their mental models of the teacher's original source movement sequence, as constructed during the teacher's live/video motor demonstration. If the decipherer's transformational outcomes were not deemed sufficient by the developers, meaning that the original source and final target movement sequences were not identical, then a fourth phase was introduced. Decipherers left the studio again to prevent communication that might affect representational refinement. Next, developers identified gaps and points of miscommunication in their original self-generated visual-graphic representation as a basis for refinement. At times, only small changes were applied to an existing representation, whereas at other times, entirely new representations were self-generated.

Phase 3–4 repetition The last two phases were repeated until the developers decided that the decipherer's performed motor interpretation (transformation) of their refined visual-graphic representation matched the teacher's original source motor sequence.

All representations were collected, photocopied, and returned to developers after each phase of development and refinement to avoid the loss of any intermediate products. Group members filed their visual-graphic products in the group portfolio that was available for examination whenever desired.

Environmental Characteristics

A learning environment's design characteristics can promote achievement of its educational goals. In our case, the goal was to promote the girls' ability for TARM through the experiences provided within the design context. In designing the current learning environment, three major setting characteristics were considered: the teacher's role, the collaborative self-generation process, and the environment's longevity. First, to maintain as much autonomy in the young girls' TARM development process as possible, the teacher was trained to cultivate students' self-chosen developmental course over the yearlong experience, by providing needed support and guidance but only minimal intervention. It should be noted that in Phase 1, the teacher was involved in instruction of the curriculum and in evaluating girls' acquired knowledge, but for the actual TARM activities during Phases 2–4, the teacher's role was minimal.

Second, a collaborative small-group setting was selected for developing TARM. Peer collaboration has shown many advantages such as: promoting thinking by producing alternative solutions for problems and multiple constructive ideas for improvement (Zhang et al. 2009); decreasing cognitive load by sharing responsibilities while working to achieve the group's goal (Pea 1993); at a practical level, maintaining the group's momentum despite participants' absence or low motivation; and from an empirical perspective, permitting explication of participants' thoughts and considerations while performing the task (Chinn and Anderson 2000). However, collaboration may sometimes constrain group work due to members' difficulties in working together and possible problematic aspects of members' social relations and in particular in cases of self-generation task.

Third, the current design incorporated repeated experiences with TARM over the long term. The environment's longevity aimed to enable gradual instruction of increasingly complex multi-dimensional and multi-element concepts (Eilam and Ofer 2016).

Results

Affordances and Constraints of the Learning Environmental Design

The present chapter focuses on a design aiming to enable a group of girls to overcome the many aforementioned challenges inherent to TARM. A very rich set of qualitative data emerged from the current research. We present visual-graphic transformational products and their accompanying discourse in terms of how they reveal the affordances and constraints of the TARM cognitive operation, the functional niche of the current learning environment including learners' spatial and temporal abilities and the social dimension.

Affording Increased Informational Accessibility

The developers' management of the represented information during their TARM processes revealed a clear emphasis on the need to improve representations' communicativeness, in line with the designed task's demands. Communicational affordances were observed both in the girls' talk about the decipherer's expected ability to interpret the representation and also during the act of drawing and writing the information representation on paper. The developers' discourse revealed many incidences of explicit verbal expressions questioning the decipherer's future ability to

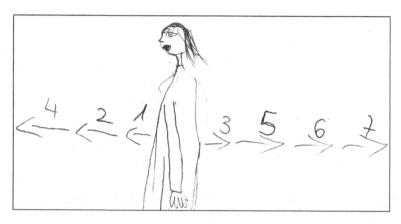

Fig. 1 Conventional symbols (ordinal numerations, arrows) and an iconic representational form (female figure in profile) used to represent repetitions of body-relative forward and backward movements

interpret the group's visual-graphic representation and transform it into her target motor one (e.g., "This is too difficult, she will be confused"). Such verbal exchanges imply an unconscious awareness of the need to make the information in the source representation accessible to the decipherer while producing the target representation. By introducing the decipherer's role as an integral part of the functional niche, the task itself encouraged communication-related considerations, which, as expected, emerged as the developers' core criteria for judging representational adequacy.

In examining these fourth-grade developers' self-generated symbolic representational language, we found dominant utilization of two main symbolic features that probably increased information accessibility: conventional symbols and iconic representations. First, the participants almost exclusively used familiar conventional symbols like arrows, letters, and numbers rather than inventing new unfamiliar symbols to represent the multi-element directional or temporal dimensions (see Fig. 1 for ordinal numerations of body-relative forward and backward movements, shown by arrows). This tendency may be attributed to the girls' attempt to enhance communicativeness by relying on shared familiar school-based exposure to conventional symbols. In some cases, variations on these conventional symbols were developed by the girls, but continued reliance on the original conventions suggests their goal of maintaining information accessibility.

Second, these young participants often selected iconic representational forms or integrated iconic symbols within their representations, such as human figures or body parts. An example is the female figure drawn in profile to show the directions of forward and backward movements as relative to the body in Fig. 1. This iconic figure may have been selected due to the spatial difficulties involved in representing the three dimensions of the human body on the "two-dimensionality of the endless flatlands of paper" (Tufte 1990, p. 12). We should note that most self-generated studies reported students' initial preferences for iconic representation (Azevedo 2000; diSessa et al. 1991; Parnafes 2012), sometimes shifting into abstract ones latter on. Iconic representations being similar to reality (i.e., the referent) increase accessibility for decipherers who do not have to interpret the more abstract forms. Our findings showed consistency in preference over time within each group with a single group preferring an abstract representation from the start.

Affording Peripheral/Tacit/Missing Aspects of Information by Emphasizing/ Complementing Whereas verbal descriptions of movements may be general (e.g., "lift your hand up"), the actual live motor performance or its graphical representation express highly specific, unambiguous, concrete parameters (e.g., raise the right hand to head level, with palm facing up and fingers pointing forward). Such transformation experiences were afforded by different intentionally designed episodes. For example, during part of the training (phase 1), the girls had to transform verbal information with tacit time dimension (source) into movement, composed of a chronological clear order of movement events enactment which represent this tacit aspect (target), as seen in Fig. 1 above.

The ordinal numeration that represents the chronology of movement exhibits explicitly the tacit temporal dimension. Another example refers to Fig. 2a and 2b below. An ambiguous information was complemented by an addition of an iconic symbol in the target graphical representation (see additional comment on next section).

Affording Information Organization

Our data analysis indicated that the developers employed two major kinds of information organization: organization on the space of the page and organization of symbols within a whole representation. The learning environment and task demands did not constrain either of these types of organization; therefore, girls could freely express graphically what they perceived of the live/video demonstrated movement by choosing the page's horizontal versus vertical layout; top-bottom, left-right, or right-left presentation of the dynamic temporal process (the represented chronology); and modes of combining distinct symbols into one whole representation. This open aspect of the design provided opportunities to consider alternatives that are usually unavailable to learners when dealing with ready-made representations and their inherent rules. However, in most cases, regarding students' organization of the target visual-graphic representation on the paper space, the dominating organizational format of presenting on the paper the represented movement event over time, were top-down or horizontal presentations. This style domination may hint at the girls' correct perception that the human movement sequences were all basically similar despite their growing complexity along the intervention, necessitating no radical changes in format.

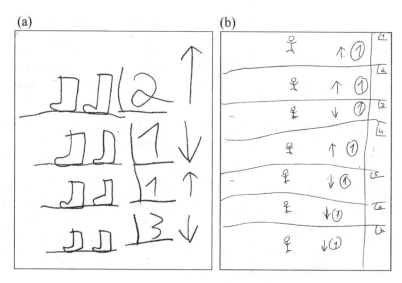

Fig. 2 A concise formula-like organization of information that was unintelligible to the decipherer (a) followed by a refined intelligible representation containing more detail and redundancies (b)

With regard to the second organizational feature, the organization of symbols within a whole representation, an interesting phenomenon emerged. Probably attempting to meet formal criteria for judging representational quality, one groups tried to create a concise "formula" to represent identical repeated movement events that they had identified within the same sequence. However, such an attempt to avoid repeated drawings of the same recurring event, while possibly achieving a more coherent, parsimonious, and compact representation (diSessa 2002, 2004), constrained the decipherer's ability to correctly transform this source visual-graphic representation into her target motor one. In this example of the formula-like representation (see Fig. 2a) the numerals represented events' number of repetitions (i.e., the numbers 2, 1, 1, 3 indicated two steps forward, one backward, one forward, and three backward). Yet, the decipherer's inability to replicate the desired sequence led to developers' refined representation (see Fig. 2b) that detailed the events chronologically in separate rows (marked with redundant ordinal numbers 1–7), thereby achieving increased efficiency in communication but eliminating their compact formula.

This example illustrates two important aspects of the intervention design experiences in TARM: the open nature of the self-generation task, which encouraged the creation and refinement of novel representations; and the feedback-driven nature of the refinement task, which triggered the girls' search for an alternative visual-graphic transformation process that would result in the decipherer's accurate interpretation. Such affordances may be expected to promote girls TARM-related considerations.

(a)   (b)

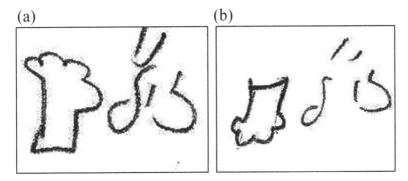

Fig. 3 An iconic image (arm pointing up or down) added to the identical alphabetic images (acronym ZY"L) for *zroa-yemina-lemaala* – right-arm-upward (a) and *zroa-yemina-lemaala* – right-arm-downward (b) to help limit the decipherer's interpretation

Affording Limitation of Interpretations

Any symbolic language may include ambiguous aspects, depending on the language specific characteristics. Such aspects require the decreasing of its ambiguity by limiting possible interpretations (Ainsworth 2006). Some products of the girls' TARM exhibited this phenomenon. In certain representations, developers used the first letters of corresponding Hebrew words to construct acronyms describing movement instruction (e.g., ZY"L for *zroa-yemina-lemaala* – right-arm-upward, as seen in Fig. 3). In their discourse about this representational form, one girl noted that confusion could ensue because in Hebrew both upward ("*lemaala*") and downward ("*lemata*") begin with the same letter Lamed, yielding the identical acronym for both. To limit the decipherer's interpretation while transforming the visual-graphical representation into a motoric one, the group decided to add an iconic image of an arm pointing either upward or downward (see Fig. 3a and b).

Next, we discuss the way our design afforded or constraint TARM with regard to the other dimensions that impacted the girls representational practices; namely, the spatial and temporal dimensions as well as the social one. Each of these dimensions of the design both affords and constrains the girls' TARM. On the one hand they afforded opportunities for practicing the mindful use of different spatial, temporal and social abilities beyond the everyday experiences, but on the other hand it required the application of challenging abilities, which were not trained previously, hence made the task highly difficult and required the girls' solutions to various related representational problems, as exemplified in the following sections.

Applying Spatial Abilities

Self-generation of visual-graphic representations of movement in space required students' application of spatial abilities. In particular, it necessitated perspective-taking and mental rotation as well as transformation of 3-D images into 2-D ones

48   B. Eilam and S. Ofer

Fig. 4 Two groups' different representations of the same sequence of spatial information. (a) Images demonstrate Group 1's use of verbal labels, indicating difficulty in presenting direction graphically; and (b) Group 4's lack of mental rotation resulting in the wrong (left) hand moving in the wrong direction. Bottom images demonstrate two different solutions to represent the 3-D movement of the lower leg backward on 2-D paper

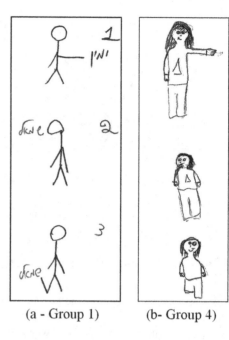

(a - Group 1)   (b- Group 4)

and vice versa. These challenges are illustrated by comparing two different groups' graphical representation of the same part of the movement sequence demonstrated by the teacher: right arm to the right, head to the left, left lower leg backward. As seen in Fig. 4a and Fig. 4b, Group 1's developers did not draw a face on their iconic human images thus had to add the Hebrew labels indicating "right" or "left" (*yemin* – right) and avoided the decipherer's need for mental rotation of the image, thus facilitating her accurate interpretation of the body.

Differently, Group 4 drew the images' faces, which determine the body-relative sides. However, while attempting to comply with the task demand to minimize verbal hints, their superior graphic solution to this problem exhibited the developers' failure to mentally rotate the direction observed in the source demonstration (see top image). This failure resulted in an incorrect specification of the left rather than right arm, pointing in the incorrect direction. The teacher's live performance and video clip of this movement, demonstrated from both front and rear views (like all others), should have activated the girls' awareness of the challenge involved in correctly representing the moving bilateral symmetrical parts of the human body.

Moreover, the bottom human images in these two groups' representations (see Fig. 4a and b) offer evidence for the students' 2-D and 3-D considerations. When asked to represent the observed live/video movement of the lower leg backward, developers in the two groups again applied two different solutions. Group 1 was troubled by their inability to represent the 3-D backward movement, as exhibited in their dialogue: "Listen, how do we draw the lower leg?" "I don't know, can she understand it?" Their solution of drawing the leg sideways (see bottom image) involved prior knowledge of anatomical joint structure and its constraint of movement, as practiced in the first phase of motor instruction: "She will understand because the leg's only possible movement is backward." Group 4 drew the moving leg as shorter to convey the perceived "disappearance" of the lower leg in this performed movement of the referent. Perhaps the developers assumed that this is an everyday familiar view that may be easily deciphered.

Applying Temporal Considerations

Participants revealed an impressive ability to find solutions to the challenges involved in time representation, including TARM involving the movement continuity and its segmentation according to a selected resolution, the chronological order of events, and simultaneity. Regarding the segmentation and as seen in Fig. 5, one group presented a dynamic whole-body movement by transforming it into a sequence of four distinct static segments, each representing a step forward or backward, in the correct chronological order.

The arrows represent the direction (i.e., up arrow indicating forward and down arrow indicating backward), and the doors constitute a redundancy by representing these same directions using the idea of open (forward) and closed (backward) doors. Although the redundancies violate the criteria of compactness and parsimony regarding the representational quality, they do emphasize the notion of distinct time units.

Regarding chronology, the girls showed awareness of its importance when transforming an ephemeral abstract source movement into a static graphic target representation. To ensure the decipherer's correct decoding of consecutive movements' chronological order, the developers added ordinal numbers indicating that they had chosen a horizontal top-down layout with right-to-left direction of presentation, as accepted for reading Hebrew (see Fig. 6).

As the intervention progressed and the movement sequences became more complex, developers encountered the need to represent multiple movement dimensions/ elements occurring simultaneously within a single time unit. Fig. 7a and Fig. 7b present two groups' examples of visual-graphic solutions to this temporal challenge. Usually, groups applied a table format with parallel columns or rows to indicate concurrent movement events, as seen in the vertical table presenting an excerpt from the sixth movement sequence (see Fig. 7a) or in the horizontal table presenting an excerpt from the ninth movement sequence (see Fig. 7b). In Fig. 7a, the left column represents the body-relative directions (forward, backward) using arrows, and the right column represents the absolute spatial directions (compass-rose numbers). Jointly, each horizontal row represents the whole body moving forward or backward in a certain direction in space. In Fig. 7b, the teacher's demonstrated live/video motor sequence was more complex, including different moving body parts in addition to body-relative and absolute spatial directions.

Thus, the girls in this group used ordinal numbers (on the right) to indicate segments' chronology, which were unnecessary due to the already familiar top-down

Fig. 5 Continuous
sequence of live/video
whole-body movements
transformed into four
discrete conventional and
iconic graphically
represented segments: step
forward, step forward, step
backward, step forward

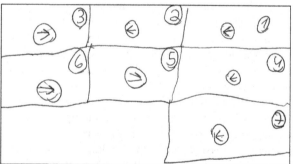

Fig. 6 Ordinal numbers added to indicate chronological order. Note the right-to-left order, as accepted in Hebrew

table format. The developers specified body parts using arm and hand icons pointed in the absolute spatial direction, as well as redundant Hebrew initials (Zayin for *zroa – arm* in rows 1 and 3; Yud for yad – *hand* in rows 2 and 3), probably to ensure identification. They gave verbal clues as to which arm/hand should be moved and in which body-relative direction. In cases of simultaneous events (e.g., row 3, involving both arm and hand movements), the word *vegam*, meaning

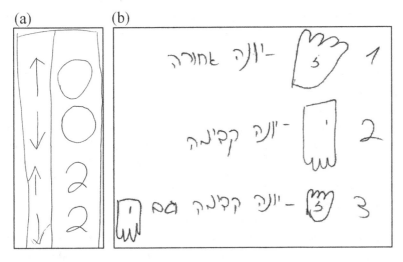

Fig. 7 Representations in table format of simultaneous movement excerpts involving several dimensions/elements (a) from Movement Sequence 6 and (b) from Movement Sequence 9

"and also", was used to indicate concurrent movements within the same time unit (see leftmost word in row 3).

Social Considerations

As found in other studies (diSessa 2004), the videos depicting the girls' TARM development processes showed that immediately after observing each live/video source movement sequence, the groups of developers spontaneously created representations with almost no discussion or reflection. Seldom did their discourse explicate the rationales for their selections or for their preferences of a specific symbol or representational form. In this sense, our collaborative design did not achieve its full potential. The rare debates about a symbol or a form did not involve discussion of representational issues but rather involved social issues, which served to constrain TARM processes. For example, a unique social phenomenon was the girls' expressed desire for the current task to provide entertainment by hiding information from the decipherer in order to create challenge and interest. At times, girls even avoided reusing previously effective symbols, decorated numbers to conceal them, or omitted a symbol because it "would be too easy" to decode. Moreover, members' collaboration within each group required compromises due to the demand to produce a joint product. Indeed, the groups emphasized the need for each member to contribute to the TARM process; therefore, some quality representations were rejected because of social arrangements: *"You don't have to make the drawings all the time... Next time it will be my turn..."* Hence, although collaboration in this complex assignment afforded a support mechanism, in some ways social issues also constrained the systematic development of TARM along the intervention.

Summary

The described environmental design that served as a "functional niche" provided unique opportunities for developing fourth-grade girls' transformational competencies. It afforded diverse transformational experiences (as one of the MRC components) that involved coping with different types of representational forms and modalities, including information – verbal-conceptual, motor, visual-graphic, ephemeral, enduring, and more – and various spatial, motoric, symbolic, and representational abilities. Those self-generated practices were based on participants' accumulated experiences with representations rather than struggle over unfamiliar representational language. This chapter showed how specific components in the environment's design afforded or constraint development of young children's TARM, specifically emphasizing the teacher's instructional but minimal guidance role, the students' collaborative setting, and the longitudinal repeated opportunities for self-generation and refinement of increasingly complex representations. In addition, the introduction of the decipherer as an evaluator of each group's representational products and enactor of live motor feedback yielded several consequences for TARM: (1) It shifted the focus of judgment from the formal quality of the representation itself (i.e., a "good representation" is compact, parsimonious, and simple) to its more limited-but-important communicative role (e.g., a "good representation" is sufficiently detailed for the decipherer to interpret, even if inefficient and containing redundancies), thereby affecting symbols' selection and organization and affecting girls' conceptualizations about effective transformations. (2) It permitted developers' identification of specific misunderstandings and/or inaccuracies in the visual-graphic product, hence facilitating its modification and promoting the girls' awareness of alternative ways for representing a particular content. (3) It enabled developers' usage of symbols and representations appropriate for peers based on shared familiar knowledge, rather than as evaluated by adults or experts. (4) However, judgments criteria promote the quality of representations as well as one's awareness of its characteristics. The girls' reliance mostly on communicational-related criteria may imply that they might have missed the opportunities provided by the self-generation experience to consider quality criteria and representational characteristics. This may be a design constraint that should be considered in future similar learning environments. Moreover, the collaborative setting also introduces some interfering processes related to the girls' desire for entertainment and demand for equal participation.

Girls' products unarguably suggested their continuous development of TARM along the time axis of the intervention, as they encountered increasingly complex movement sequences. However, students already entered the intervention with rich representational resources, albeit mostly informal and inexplicit (Sherin 2000);

therefore, we cannot pinpoint the exact contribution of the current formal, systematic, explicit set of experiences to their TARM knowledge development and refinement. Moreover, through their transformations and representational products, these fourth graders displayed understanding of human movement, suggesting that alongside TARM development they also developed their conceptual understanding of the domain represented. In contrast, the need to consider abstract temporal and spatial aspects, which requires high-order thinking, still remained a challenge throughout the entire intervention. This was evidenced in both erroneous graphical solutions and more simplistic organizations that never reached a formal high quality of representation despite its correctness. Future development of the current intervention design should consider the introduction of scaffolds (i.e., support provided by a teacher or knowledgeable others, a computer or a paper-based tool that enable students to meaningfully participate in and gain skill and knowledge from a task that they would be unable to complete unaided; Belland 2014).

In sum, the described learning environment may serve as a model for other planning interventions to develop diverse aspects of young learners' TARM competence and domain knowledge. Such environments may be designed according to specific contents and representational goals in order to control for different levels of task complexity and different TARM skills (e.g., representing a static phenomenon does not require temporal considerations). We claim that the current environment's affordances will remain across participants, contexts, situations, and tasks. We encourage teachers to integrate such TARM tasks directly into their classrooms and look forward to future studies that use this design or modify it to achieve particular objectives.

References

Ainsworth, S. E. (2006). DeFT: A conceptual framework for considering learning with multiple representations. *Learning and Instruction, 16*, 183–198.

Azevedo, F. S. (2000). Designing representations of terrain: A study in meta-representational competence. *Journal of Mathematical Behavior, 19*, 443–480.

Bamberger, J. (2007). Restructuring conceptual intuitions through invented notations: From path-making to map-making. In E. Teubal, J. Dockrell, & L. Tolchinsky (Eds.), *Notational knowledge* (pp. 81–112). Rotterdam: Sense.

Belland, B. R. (2014). Scaffolding: Definition, current debates, and future directions. In J. M. Spector (Ed.), *Handbook of research on educational communications and technology*. .(Chap. 39 (pp. 505–518). New York: Springer.

Bertin, J. (2007). *Semiology of graphics: Diagrams, networks, maps. (original 1983, translated by berg, W.)*. Madison, WI: University of Wisconsin Press.

Chinn, C. A., & Anderson, R. C. (2000). The structure of discussions that promote reasoning. *Teachers College Record, 100*, 315–368.

Eilam, B. (2012). *Teaching, learning, and visual literacy: The dual role of visual representation in the teaching profession*. New-York, NY: Cambridge University Press.

Eilam, B., & Gilbert, J. (2014). The significance of visual representations in the teaching of science. In B. Eilam & J. Gilbert (Eds.), *Science teachers use of visual representations, Series of Models and modeling in science education: Springer.* (pp. 3–28). Switzerland: Springer.

Eilam, B., & Ofer, S. (2016). Meta-representational competence: Self-generating representations of human movement. *Manuscript in advance preparation.*

Eshkol, N., & Wachman, A. (1958). *Movement notation.* London: Weidenfeld and Nicholson.

Gallahue, D., & Ozmun, J. (1998). *Understanding motor development: Infants, children adolescents, adults* (5th ed.). New York: McGraw Hill..

Hammill, D., Pearson, N. A., & Voress, J. K. (1993). *Developmental test of visual perception – Second edition (DTVP2).* Austin, TX: Pro-Ed.

Kozma, R. (2003). The material features of multiple representations and their cognitive and social affordances for science understanding. *Learning and Instruction, 13,* 205–226.

Lehrer, R., Schauble, L., Carpenter, S., & Penner, D. (2000). The interrelated development of inscriptions and conceptual understanding. In P. Cobb, E. Yackel, & McClain (Eds.), *Symbolizing and communicating in mathematics classroom: Perspective on discourse, tools, and instructional design* (pp. 325–360). Mawah, NJ: Laurence Erlbaum associates, Inc.

Ofer, S. (2001). Movement literacy: *Development of the concept and its implications for curriculum* (Unpublished masters' thesis). University of Haifa, Israel (Hebrew with English abstract).

Ofer, S. (2009). *Development of symbolic language to represent movement among fourth graders* (Unpublished doctoral dissertation). University of Haifa, Israel (Hebrew with English abstract).

Parnafes, O. (2012). Developing explanations and developing understanding: Students explain the phases of the moon using visual representations. *Cognition and Instruction, 30,* 359–403.

Pea, R. D. (1993). Practices of distributed intelligence and designs for education. In G. Salomon (Ed.), *Distributed cognitions: Psychological and educational considerations* (pp. 47–87). New York: Cambridge University Press.

Pluta, W. A., Chinn, C. A., & Duncan, R. G. (2011). Learners epistemic criteria for good scientific models. *Journal of Research in Science Teaching, 48,* 486–511.

diSessa, A. A. (2002). Students criteria for rep adequacy. In K. Gravemeijer, R. Lehrer, B. van Oer, & L. Verschaffel (Eds.), *Symbolizing, modeling and tool use in mathematics education* (pp. 105–129). Dordrecht, The Netherlands: Kluwer.

diSessa, A. A. (2004). Meta representation: Native competence and targets for instruction. *Cognition and Instruction, 22,* 293–331.

diSessa, A. A., & Sherin, B. L. (2000). Meta representation: An introduction. *Journal of Mathematical Behavior, 19,* 385–398.

diSessa, A. A., Hammer, D., Sherin, B., & Kolpakowski, T. (1991). Inventing graphing: Meta-representational expertise in children. *Journal of Mathematical Behavior, 10,* 117–160.

Seufert, T. (2003). Supporting coherence formation in learning from multiple representations. *Learning and Instruction, 13,* 227–237.

Sherin, B. (2000). How students invent representations of motion. *Journal of Mathematic Behavior, 19,* 399–441.

Transform. (2014). In *Merriam-Webster's* online dictionary. Retrieved from http://www.merriam-webster.com/dictionary/transform

Tufte, E. R. (1990). *Envisioning information.* Cheshire, CT: Graphics Press.

Tufte, E. R. (1997). *Visual explanations.* Cheshire, CT: Graphics Press.

Tversky, B. (2005). Functional significance of visuospatial representations. In P. Shah & A. Miyake (Eds.), *The Cambridge handbook of visuospatial thinking* (pp. 1–34). Cambridge: Cambridge University Press.

Van Meter, P. (2001). Drawing construction as a strategy for learning from text. *Journal of Educational Psychology, 93,* 129–140.

Verschaffel, L., Reybrouck, M., Jans, C., & Van Dooren, W. (2010). Children's criteria for representational adequacy in the perception of simple sonic stimuli. *Cognition and Instruction, 28,* 475–502.

Zhang, Z. H., & Linn, M. (2011). Can generating representations enhance learning with dynamic visualization? *Journal of Research in Science Teaching, 48,* 1177–1198.

Zhang, J., Scardamalia, M., Reeve, R., & Messina, R. (2009). Designs for collective cognitive responsibility in knowledge-building communities. *Journal of the Learning Sciences, 18,* 7–44.

Agreeing to Disagree: Students Negotiating Visual Ambiguity Through Scientific Argumentation

Camillia Matuk

Introduction

Goals of Scientific Argumentation

Argumentation is the process by which partners work to achieve consensus over a shared explanation (Sandoval and Millwood 2005). It is also at the core of scientific practice (Driver et al. 2000). Scientists argue about such things as the interpretation of data, the rationale behind an experimental design, and the reasoning behind a theoretical perspective (Latour and Woolgar 2013). Their explanations are accounts that establish logical relationships between claims and evidence in order to specify the causes for observed effects (National Research Council 1996). Being fluent in the use of evidence is therefore important for successfully engaging in argumentation.

Individuals can achieve consensus through argumentation by two means (Fischer et al. 2002). One is by integrating their differing views to create a complete explanation. Another is for one perspective to outcompete the other. As partners work to co-construct an explanation through argumentation, they move between the interconnected activities of sensemaking, articulation, and persuasion (Berland and Reiser 2009). However, when partners' efforts to persuade each other of their perspectives fail, a third possibility is that no consensus is achieved. Such failed arguments can occur for multiple reasons. For example, partners may misunderstand each other's arguments; they may misunderstand the topic; they may have difficulty articulating the connection between evidence and their claims (McNeill and Krajcik 2007; McNeill et al. 2006); or they may resist accepting another's point of view even though they recognize its logic. Another reason is that partners may disagree

C. Matuk (✉)
New York University, New York, USA
e-mail: cmatuk@nyu.edu

© Springer International Publishing AG, part of Springer Nature 2018
K. L. Daniel (ed.), *Towards a Framework for Representational Competence in Science Education*, Models and Modeling in Science Education 11,
https://doi.org/10.1007/978-3-319-89945-9_4

on the evidence that is the basis of each other's arguments. The latter, I maintain, is a problem particularly associated with ambiguous visual evidence.

This chapter examines the implications of ambiguity in graphs during scientific argumentation, and the skills necessary for handling it. I first offer two examples of ambiguous graphs in scientific argumentation: one in the context of the Space Shuttle Challenger disaster, and another in the context of global climate change. I then examine an episode of two middle school students' disagreement over the meaning of a graph during a technology-enhanced inquiry unit on global climate change. Through this example, I illustrate ambiguity as constructed by students both from their individual prior knowledge and expectations, and through their discourse. Analysis of this case highlights opportunities for learning to argue when instruction acknowledges ambiguity and legitimizes disagreement.

Data, Evidence, and Interpretation: The Objects of Argumentation

There are different perspectives on the relationship between data and evidence, and their role in argumentation. Some maintain that data can indicate whether or not a theory may be correct (Mulkay 1979), a view of data as having a fixed, inherent meaning. Others consider theories to be never conclusively proved or disproved (Duhem and Quine, cited in Grunbaum 1960). In this view, data is transformed through negotiated interpretations into evidence to support or refute a theory (Amann and Knorr Cetina 1988). Data is thus inseparable from theoretical controversies (Collins 1998). Sometimes, the same data may be used as evidence to support different views, as in the public debate over global climate change, discussed further below. In such cases where consensus through argumentation is not achieved, evidence may be described as ambiguous.

What Causes Visual Ambiguity?

An ambiguous message, either verbal or visual, is one that lacks clarity in meaning, whether or not intentionally (Empson 1932). For instance, some messages are ambiguous because they were designed to have more than one meaning (e.g., Fig. 1). Others have gained new meanings over time. The swastika, for instance, once widely recognized as an Eastern symbol of good fortune, now tends to be regarded as a symbol of Germany's Nazi party. Other messages are ambiguous because they are likely to be interpreted differently. For instance, the sentence "Place the bell on the bike in the garage" might mean that the bike already has a bell and is to be placed in the garage, or that the bell is to be placed on the bike, which is in the garage.

Fig. 1 *"Landscape shaped like a face*. State 1." by Wenceslaus Hollar (1607–1677), scanned by the University of Toronto from the University of Toronto Wenceslaus Hollar Digital Collection

It has been argued that visual messages are particularly likely to be ambiguous. For instance, Barthes (1977) describes images as polysemic in that they connote various possible meanings. How viewers choose to focus on some meanings while ignoring others depends not only on the structure of the image, but also on the image's context, and on the viewer's personal characteristics, including their prior knowledge and expectations (Shah et al. 2005). In this sense, images do not so much hold inherent meaning as they are sites to which viewers can attach meaning (Edwards and Winkler 1997). In the context of argumentation, this meaning is co-constructed through partners' interactions.

Ambiguity in Graphs

Graphs are one kind of visual that is commonly used as evidence in science, both in traditional print media (Kaput 1987; Lewandowsky and Spence 1989; Mayer 1993; Zacks et al. 2002) and in technology-based science learning environments (Nachmias and Linn 1987; Quintana et al. 1999; Reiser et al. 2001; Scardamalia et al. 1994). A reason for the widespread use of graphs is in their ability to clearly depict quantitative patterns (MacDonald-Ross 1977; Tversky 2002; Winn 1987) in ways that take advantage of the human capacity for spatial information processing (Kosslyn 1989). Yet, as with other kinds of visuals, the meaning of graphs can be ambiguous to viewers, and numerous examples from prior research illustrate how students misinterpret them. As Shah and Hoeffner (2002) review, a graph's interpretation depends on its visual characteristics, on viewers' knowledge about the graph's symbolic conventions, and on the alignment between the graph's content and viewers' prior knowledge and expectations of that content.

Kinds of Ambiguity

There are different reasons for ambiguity. Futrelle (2000) identifies *lexical ambiguity* as occurring because the context is insufficient to determine which of the several possible meanings a message is intended to communicate; and *structural ambiguity* as occurring because the configuration of the message conveys more than one meaning. Further kinds of ambiguity can be distinguished by their sources. As Eppler et al. (2008) describe, the visual can be a source of ambiguity in terms of the designer's choice of icons, symbols and metaphors (i.e., iconic, symbolic, and indexical ambiguity). The people interpreting the visual can create ambiguity by the different professional and cultural backgrounds they bring to their interpretations (i.e., background ambiguity. See also Gaver et al. 2003; and d'Ulizia et al. 2008). Finally, collaborative interactions between interpreters can cause ambiguity, such as when each unknowingly focuses on different parts of the same visual (i.e., focus ambiguity), or assumes different goals of the same visual (i.e., scope ambiguity).

The Space Shuttle Challenger Disaster and the Global Warming Hiatus

Prior research discusses ambiguity as a challenge in individual sense making (e.g., Nemirovsky and Noble 1997), as a frustration among collaborators (d'Ulizia et al. 2008, Avola et al. 2007), and even as a cause of catastrophic events (Tufte 1997). Debate over the Space Shuttle Challenger disaster, for example, has focused on the possibility of an ambiguous chart leading to a misinformed decision to launch the shuttle, which disintegrated less than two minutes into its flight, and left its seven crew members dead. As Tufte (1997) claims, NASA would have decided to cancel the launch had the engineers more clearly represented the precarious relationship between cooling temperatures and the degree of damage to the O-ring seal on the shuttle's solid rocket booster joint (Fig. 2). Robison et al. (2002), however, explain that the graph Tufte created to demonstrate his argument (Fig. 3) is itself flawed. It plots data that was not available to engineers, and also displays the wrong data, failing to distinguish between O-ring and ambient air temperatures. While Tufte's graph succeeds in supporting his argument, it fails to display the effect that was relevant for making an informed decision about the shuttle launch.

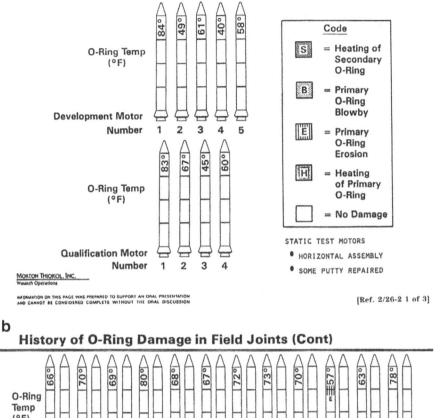

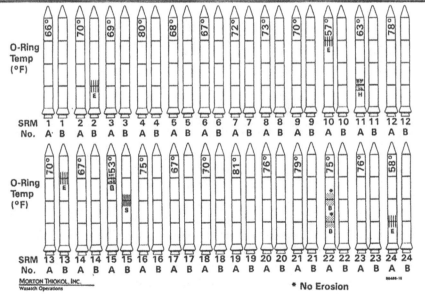

Fig. 2 History of O-Ring Damage in Field Joints. From the Report of the Presidential Commission on the Space Shuttle Challenger Accident, February 26, 1986 Session. Vol. 5, pp. 895–896. (http://history.nasa.gov/rogersrep/genindex.htm)

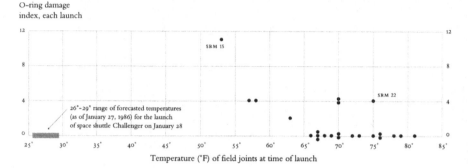

Fig. 3 Tufte's graph showing the relationship between temperature and O-ring damage on the Space Shuttle Challenger. Courtesy of Graphics Press, LLC

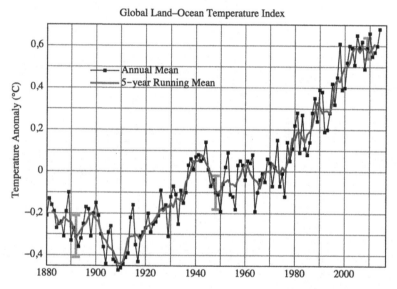

Fig. 4 Mean global surface temperatures from 1880–2014. The annual mean is shown by the black line, and the 5-year running mean is shown by the red line. Uncertainty estimates are shown by the green bars. From NASA Goddard Institute for Space Studies (http://data.giss.nasa.gov/gistemp/graphs/)

Another example of ambiguous representation is in the debate over the existence of an apparent global warming hiatus: a pause since 1998 in the trend of the Earth's rising surface temperatures. Arguments for and against a hiatus have pointed to the same data to support vastly different narratives (e.g., Carter 2006; Karl et al. 2015). In effect, the appearance of an increasing or decreasing trend is sensitive to the start date along the graph's x-axis. That is, beginning in 1996, temperatures appear to increase by 0.14 °C, while beginning in 1997, they appear to increase by only 0.07 °C per decade (Stocker et al. 2013). The scale and variation of the entire set of temperature readings (Fig. 4) make it easy to cherry pick data to support either opposing view (Easterling and Wehner 2009). For skeptics, graphs of these cherry-picked data have been persuasive evidence for confirming their beliefs, and for persuading others to deny the fact of climate change. Altogether, people's expectations, the capacity for ambiguity in temperature graphs, scientists' poor communication, and the media's failure in balanced science reporting, have led to unproductive counter narratives in the public debate over climate science (Mooney 2013).

Examples such as these illustrate how ambiguity can arise from the interaction between factors influencing the interpretation of visual representations. Differences in viewers' expert knowledge of the content, their understanding of the context and of the goals of the graph, and their expectations of the message, can all influence how viewers interpret a graph, as well as how they use it as evidence to persuade others of their interpretations.

Uses of Ambiguity

The literature tends to refer to ambiguity in terms of its negative consequences. As illustrated by the examples above, inconsistencies in participants' expectations and knowledge, paired with vaguely depicted information, can lead to misaligned interpretations. These can have potentially disastrous outcomes, or at the very least, create frustration when participants are unable to resolve differing interpretations of the evidence.

However, ambiguity can also be a resource that enriches and advances collaboration. By allowing multiple perspectives to coexist, ambiguity can be conducive to collaboration between diverse individuals in organizations, enable flexibility in adverse conditions, and thus ensure greater stability (Eisenberg 1984). Among designers, ambiguity can spur discussion of new ideas and encourage consideration of alternative perspectives (Eppler and Sukowski 2000; Eppler et al. 2008). As will be discussed here, ambiguity can even promote productive learning behaviors within collaborative learning settings. That is, through their attempts to persuade one another of their own interpretations, partners will work to articulate their points of view. Doing so can help expose gaps in their knowledge, and develop their abilities to critique their own and others' perspectives.

Prior research considers the skills necessary to interpret graphs among both students (Friel et al. 2001) and professional scientists (Amann and Knorr Cetina 1988). Some research examines how verbal ambiguities can lead to misunderstandings regarding the interpretation of graphs (Bowen et al. 1999). However, research in science education tends to examine the interpretation of graphs designed to have single, correct meanings. Indeed, graphs within learning environments are often designed to be interpreted as such, especially at the middle school level. As a result, there are few opportunities for students to encounter legitimate visual ambiguities, nor for researchers to understand the skills needed to resolve misunderstandings over ambiguities, and to reap their benefits.

This research considers a case of two students' differing interpretations of the same graph, and the consequences of their efforts to build consensus through argumentation. Through a close examination of students' rhetorical uses of the graph, I describe emerging skills in sensemaking, articulation, and persuasion in the context of using ambiguous visual evidence (Berland and Reiser 2009).

Case Background: The Global Climate Change Unit

The case described below draws on data collected during the enactment of a 5-day long computer-supported classroom curriculum unit called Global Climate Change (GCC) developed in the Web-based Inquiry Science Environment (WISE, wise. berkeley.edu). Research finds middle and high school students' ideas about global climate change to be both simplistic and narrow in scope (Shepardson et al. 2009). Students hold rudimentary mental models of the earth's climate system, and fail to articulate any complexity underlying the causes and impacts of global warming, and of its potential resolutions. The WISE GCC unit was designed to introduce middle school students to the notion of global climate change as a complex process with various causes (Rye et al. 1997).

Following the Knowledge Integration framework (Linn et al. 2004), the unit's activities first elicited students' initial ideas on global climate change, added to students' repertoire of ideas, encouraged them to develop criteria in order to sort and distinguish these various ideas, and then guided them to refine the connections among ideas toward developing a normative understanding of the topic. Embedded prompts guided students' inspection of graphs and their explorations of NetLogo simulations (ccl.northwestern.edu/netlogo) as they attempted to understand how solar radiation interacts with the surface and atmosphere of the earth, and the associated impacts of human activity on levels of greenhouse gases (Svihla and Linn 2012).

GCC was completed by 55 pairs of middle school students taught by the same teacher. Student pairs worked self-paced through the unit as their teacher circulated and offered guidance as needed. The case described below focuses on two partner students, referred to here as Tad and Kino, and chosen because of their vocal commitments to opposing interpretations of a graph. While atypical among their classmates, Tad and Kino provide a good illustration of potential for rich and productive discourse around ambiguous visuals in students' scientific argumentation. I offer an in-depth description and analysis (Schoenfeld et al. 1991) of the discourse between Tad and Kino that identifies and illustrates some distinct rhetorical moves by Tad and Kino for persuading one another of their conflicting viewpoints. I consider the skills apparent in their use of the graph as evidence in their arguments, and how, in the manners by which they attempt and fail to persuade one another of their interpretations, certain emergent skills for handling ambiguity in visual evidence become clear.

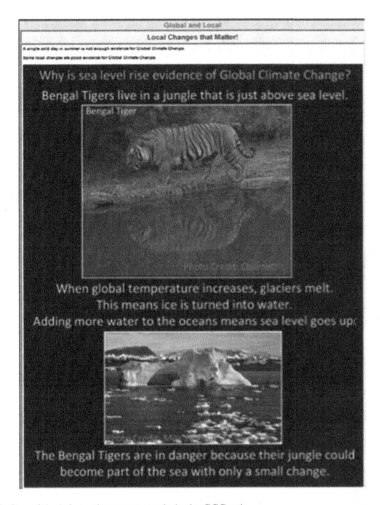

Fig. 5 One of the information screens early in the GCC unit

"How has global climate changed?"

The unit begins by explaining that there were numerous glacial periods in Earth's history, and that melting ice and rising polar sea levels are evidence of increasing global temperatures (Fig. 5). Following this, the unit shows a graph that plots global temperature over time (Fig. 6). Created by the unit's designers based on publically available temperature data, the graph's x-axis extends from the formation of the Earth to the present, and is marked with five glacial periods. The temperature curve begins at the appearance of the first life on Earth, and fluctuates until the present day.

a

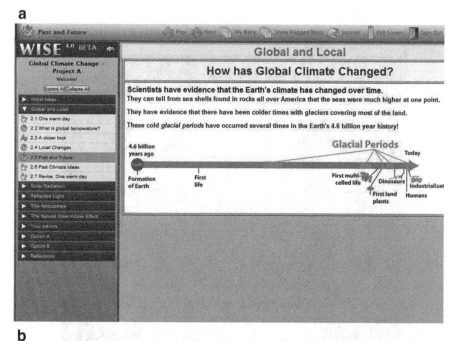

b

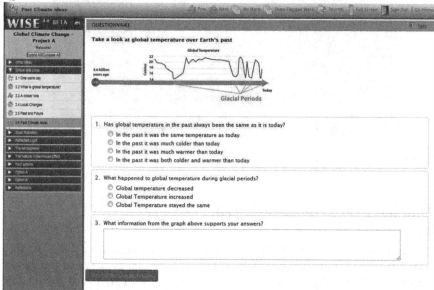

Fig. 6 Two sequential screens from the GCC unit showing the graphs in Tad and Kino's dispute

In the case examined, Tad and Kino are discussing the following three questions, the first two of which asked them to select one of the multiple choice responses.

Take a look at global temperature over Earth's past.

1. Has global temperature in the past always been the same as it is today?

- In the past it was the same temperature as today
- In the past it was much colder than today
- In the past it was much warmer than today
- In the past it was both colder and warmer than today

2. What happened to global temperature during glacial periods?

- Global temperature decreased
- Global Temperature increased
- Global Temperature stayed the same

3. What information from the graph above supports your answers?

Below is an excerpt of Tad and Kino's discussion of the first question. In it, Tad expresses favor for the fourth choice, "In the past it was both colder and warmer than today." Meanwhile, Kino believes that "In the past it was much colder than today." I elaborate on the reasons for each of Tad's and Kino's interpretations, and describe their attempts to convince each other of their views.

Two Interpretations of the Same Graph

Throughout the unit, Kino maintained the notion that global temperature was rising. Although he was not incorrect in his belief that global temperatures were rising over time, Kino's commitment to this idea was to the extent that he had reduced the idea to the phrase, "Global warming made it warmer." He would utter this at the end of each activity as though to confirm it to himself and to Tad that this was the main takeaway (Fig. 7).

a

b

Fig. 7 Kino (*right*) indicating to Tad (*left*) that "Global warming made it warmer"

This view thereafter framed Kino's interpretation of the graph. Rather than acknowledge the variations in the global temperature curve over time, as Tad did, Kino attended only to those parts of the graph that confirmed the trend of rising global temperatures. The most salient evidence to him was not the ups and downs of the temperature curve, but the blue bars along the green x-axis, which marked the glacial periods of Earth's history. That each blue bar in the series was shorter in length than the last meant that ice ages were becoming shorter over time. Kino saw this as a trend that was consistent with the evidence presented earlier in the unit, that polar ice caps were melting and causing sea levels to rise.

This view was not necessarily incorrect, but rather one that generalized the smaller scale temperature variations in Earth's history. The designers of this graph activity intended for students to notice that temperature *did* fluctuate over the course of Earth's history, in spite of its general increase over time. Other markings, such as the blue bars indicating the ice ages, were merely there to put these fluctuations into historical context. But as Kino sought evidence to confirm his view, these markings became a source of ambiguity in his discussion with Tad.

In the following excerpt, Kino presents the reasoning behind his claim that global temperature was colder than it was today, and points to the shortening lengths of the blue bars as evidence.

K: So, say in the past, was it colder than today? So, does it, it means like was this colder than this? [with one hand, points to the longest ice age near the beginning of the time line, and with the other, to the shortest ice age near the present.] No, 'cause that could be like, look at the difference. This is huge.

T: But it [i.e., the question] says 'in the past.' 'Has global temperature in the past always been the same as today?'

K: No, it's colder than, it's colder than today.

T: Well look, today, that was about 10 years ago [uses the cursor to indicate the curve toward the right end of the x-axis] right now we'd be up here [indicates a high point along the curve closest to the right of x-axis]...

K: [interrupts] So you're saying this [indicates the left end of the x-axis] is not colder than this [indicates the right end of the x-axis]?

T: No, it's longer [i.e., the earliest glacial period is longer than the later one]. But we didn't have anything to do with that.

K: Yeah, longer.

T: Yeah but we didn't have to do with anything to do with the shortening of the ice ages. Nothing. Civilization was there [points to the very end of the x-axis, after the final blue bar that marks the most recent ice age.]

K: Yeah! [spoken in a tone of impatience]

Kino appeared exasperated at failing to convince Tad of the relevance of the blue bars. He pointed out the differences in length between two of them as evidence that the Earth is experiencing shorter periods of cold. To this, Tad offered two rebuttals. First, he stated that while the earlier ice ages were longer than the later ones, this did not mean that they were colder. Second, he pointed out that civilization only began

Fig. 8 Tad (*left*) leading Kino (*right*) through his reasoning behind using the temperature as evidence

after the ice ages ended, implying that, given the role of human activity in global warming, temperature evidence before civilization was not relevant in explaining the pattern of increasing temperature that Kino sought. Kino acknowledged neither of these rebuttals, and as he and Tad turned their attention back to the prompt, both reiterate their opposing interpretations:

T: So, was it warmer, or --- In the past, in the past. Was it about the same temperature?
K: [mumbles]
T: About the same?
K: [mumbles]
T: Colder and warmer?
K: I think it's colder.
T: I think it's warmer.

Tad's next attempt was to similarly walk Kino through his reasoning Fig. 8. Although his line of questioning did lead Kino to utter Tad's view that global temperature in the past was at different points both warmer and colder than today, Kino continued to resist it.

T: Just answer my questions, OK?
K: Mm.
T: Yes or no. Was it hotter than today?
K: Yeah.
T: Was it colder than today?
K: What?
T: Was it colder than today?
K: Yeah.
T: There's your answer.
K: It's colder than today?
T: You said it.
K: Was it colder than today?
T: Was it hotter than today, too?
K: Yeah, so it's getting warmer. And warmer and warmer.

Inferring Evidence

Having earlier similarly failed to get Tad to accept his interpretation of the blue bars, Kino next attempted to convince Tad that even the temperature curve supported his view. Not limiting himself to the visible evidence, he explains how extrapolating the temperature curve tells a story of Earth's increasing temperature:

K: So if this was hot, uh, let's say, um, a million years ago [indicates the high points along the curve], it would be a lot hotter, a lot hotter like 20, 20 million years-- Like it was super hot here [uses the cursor to indicate another high point on the curve just after the first glacial period]. So I think it's going to be even hotter here [moves the cursor to the very end of the curve]

T: But you're only saying that because right now, you're thinking that us is right here [indicates a space above the imaginary point, to which Kino had earlier extrapolated]. They're asking us about that [broadly indicates the entire graph].

K: Dude, you're not using this part [uses finger to indicate the green x-axis].

T: Yes I am.

K: No, you're not. If you're using it [points to each of the blue strips marking the glacial periods along the x-axis] then this would be getting smaller every like, every like, say like a hundred years. (Fig. 9).

Tad's response was to remind Kino that their reasons should not be based on inferences, but on the visible evidence available. To this, Kino pointed out that Tad himself was not using the visible evidence, as he neglected the blue bars on the x-axis.

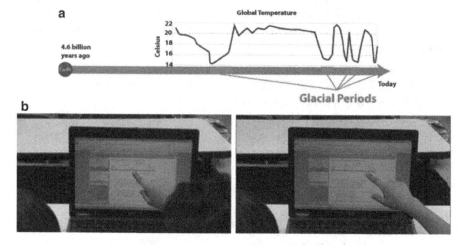

Fig. 9 (a) The graph under dispute. (b) Kino pointing to the blue bars on the x-axis

Reading the Designer's Mind

Having failed several times to convince Tad of the validity of his evidence, Kino next proposed that his claim was likely the one they were expected to give. It was possible, he reasoned, to guess which evidence they were intended to use by the design of the activity:

K: Listen, listen. If this thing is so important [indicates the temperature curve in Fig. 4, bottom], they [i.e., the unit's designers] would have put it over here and gave some information [navigates to the previous screen (Fig. 4, top)]. That's the only important thing [i.e., the x-axis]. That's why they only put, that's why they only put this.

Kino thus demonstrated an awareness of the task as a designed activity. He saw the questions and each element of the visual display to be the result of someone's conscious decisions. As he explained to Tad, their goal as students was to guess these intentions, and to respond accordingly.

Reaching Consensus Through Compromise

It was clear that both students were anxious to move on from this question given the little class time that remained. Agreeing that it was more important to complete the unit on time than it was to come to consensus over this question, Tad eventually turned his attention away from Kino and toward the laptop:

T: I'm just going to write.
K: Yeah, but write 'I.'
T: [types "We got…"]
K: [reaches over to the keyboard and hits the backspace to delete what Tad had typed] Dude, write "I." You do it the wrong way. I don't care.
T: [urges Kino to calm down].
K: Just do it your way. I don't care if we got it right.

While Kino was resigned to the fact that consensus would only be reached by agreeing to disagree, he still felt strongly about the validity of his argument. To compromise, Kino would permit Tad's answer to be the official one the two would submit, but his consolation would be in distinguishing it as Tad's rather than as their shared idea.

It is notable that Kino's desire to not be misrepresented (requesting that Tad type "I" instead of "we") was stronger than his desire to submit what he felt was a correct response. When Tad made further attempts to persuade Kino of his view, Kino impatiently urged him to simply type his answer so that they could both proceed to the next question.

K: Dude, we're still on this question. Just do it.

T: Can you calm down? Can you calm down and talk like humans?

K: [stands abruptly and walks away]

Kino was visibly agitated, but soon returned to his seat next to Tad. Perhaps as a conciliatory measure, Tad offered to allow Kino to give his response to the next question, since Tad gave his own to the previous one. He seemed prepared to accept that although they disagreed, both of their ideas might be correct:

T: For this one, we're going to agree. 'What information from the graph above supports your answer?' For your answer, if we put your answer right there, what information did you use? The same information as I used for this one [refers to the previous question]?

K: No. I used this [indicates the green x-axis].

T: And I used that, too. We used the same things here. 'What information from the graph above supports your answers?'

K: That thing [indicates the green x-axis]. I don't know what to call that thing. Not the graph [covers the temperature curve with the palm of one hand]. I don't need the graph. This thing [points to the green x-axis visible below his hand]. (Fig. 10)

T: But that's the point. You're not getting it.

K: What?

T: You should use the graph. That's, that's –

K: I used the graph. But the thing that supported my answer was this. Because it supported, not used. I used this [indicates the entire graph] but this [indicates green axis] is the thing that supported my answer.

T: [types] "We used the ice age line to support our answers and we used the timeline of the page before." "Good enough?"

K: [agrees]

In this final excerpt, Kino is most explicit in his admission of being selective in his choice of evidence. As he described, he referred generally to the graph, but chose to focus on the x-axis as evidence to support his claim. Tad, conceding this question's answer to Kino, types and submits it without retort.

Fig. 10 Kino covers the temperature curve and points to the x-axis to demonstrate to Tad the portion of the graph that he considers to be evidence for his claim

Conclusions and Implications

Tad and Kino's discussion illustrates how ambiguity in visual, scientific evidence is created from viewers' expectations and from the structure of the visual. Each element of the graph was available as evidence to support different interpretations, and each became more or less salient as evidence depending on the viewer's expectations.

It became clear early on that the disagreement between Tad and Kino was the result of what Eppler et al. (2008) call *focus ambiguity*: Each based their interpretations on different portions of the graph. Their disagreement may also be described as a result of *scope ambiguity,* by which viewers assume different purposes of the same visual (Eppler et al. 2008). Kino's strong expectations of the graph to confirm Earth's rising temperatures, and his assumption that this was the response the unit's designers intended for students to give, shaped the portion of the graph to which he attended. These expectations also prevented Kino from acknowledging any relevance of temperature curve, and thus, of Tad's interpretation. Likewise, Tad's literal approach to the question prevented him from seeing the validity of Kino's argument, and prompted him to ignore the lengths of the ice ages as an alternative way to interpret the graph.

This case demonstrates how ambiguity can arise from the discursive interaction between partners. Had Tad and Kino been working independently, neither would have challenged their individual interpretations, and the possibility of the graph's different potential meanings would not have arisen. Focus ambiguity and scope ambiguity can both be positive when they lead partners to realize new possible interpretations of the visual, and new uses for it. But they can also be negative when they result in confusion and misunderstanding (Eppler et al. 2008).

Skills in Persuasion with Visually Ambiguous Evidence

Although Tad and Kino experienced frustration over their inabilities to resolve the graph's ambiguities, the process of their argument, through which they attempted to reach consensus, ultimately elicited productive learning behaviors that other research suggests can give rise to conceptual learning (Nussbaum 2008). Specifically, the graph's ambiguity encouraged both students to explicitly articulate the connection between claims and evidence, which is an important skill in scientific argumentation (Berland and Reiser 2009). They also exhibit general skills in graph comprehension, including the ability to recognize features of a graph, to explain relationships, and to contextualize these in the discipline (Shah and Hoeffner 2002). In their attempts to persuade one another, they coordinate these and other skills in ways one might consider particular to negotiating visual ambiguity. I outline these skills below.

Making Explicit Visual References to Coordinate Claim and Evidence

One way to resolve focus ambiguity is by making clear what is being referred to, and a simple way to do this is by pointing (Eppler et al. 2008). While both Tad and Kino used gesture to indicate the components of the graph they used as evidence for their competing claims, they could not agree on which was most relevant. They therefore walked through the reasoning behind their arguments by pointing to specific locations of the graph, and narrating their connection to their claims. Doing so involved the ability to chunk the graph into meaningful components (e.g., distinguishing the temperature curve from the x-axis); to think analytically about the meanings of these components in relation to the others (e.g., conceiving of temperature in terms of small scale changes vs. general trends), and selecting persuasive comparisons (e.g., the highest and the lowest points of the temperature curve, or the longest and shortest bars on the x-axis).

Re-Interpreting Evidence to Reinforce an Argument

The failure of one strategy to persuade prompted both Tad and Kino to seek alternatives. Each student first made qualitative comparisons: Tad showed Kino that there were both lower and higher points on the curve compared to the present; and Kino showed Tad that the blue bars in the past were longer than they were in the present. When these strategies failed to convince, Kino translated his visual comparisons into numeric ones, pointing out that the ice ages were shortening every hundred years. His ability to re-interpret the graph, moving between qualitative and quantitative descriptions of the data, provided him with new, potentially more convincing ways to support his original view.

Making Inferences to Emphasize the Significance of an Observed Pattern

Unable to convince Tad by reference to the available evidence, Kino extrapolated it. He predicted temperatures would be even higher beyond the timeline shown as though to emphasize the significance of the observable trend. Kino's ability to make this inference required a knowledge of the kind of reasoning graphs allow, which is not only a core graphing skill (Glazer 2011), but also one that can help communicate and persuade one of the meaning of a pattern.

Reasoning through Connection between Data and Context

On understanding that Kino was basing his claim of increasing global temperatures on the shortening ice ages, Tad refocused their discussion within the context of the data. He pointed out that civilization only began after the ice ages. Thus, given the role of humans in global warming, temperature data as far back as the ice ages was not relevant. This required an ability to see the data not just for what it is, but to also interpret it within its disciplinary context.

Understanding a Visual's Multiple Purposes

Kino final attempt to persuade Tad demonstrated his awareness of the graph having been designed for more than one purpose. That is, he realized it was not just as a scientific representation, but also a representation within a learning environment. Based on what was shown and not shown in the unit's previous screens, he guessed that the unit's designer expected a certain response from them as students. Given the lead-up to that point in the unit, it was likely that the expected response was that global temperatures were colder than in the past. Kino's awareness of a designer became yet another source of evidence to support his interpretation, as well as a means to persuade his partner.

Designing to Leverage Visual Ambiguity

Supporting students' meaningful engagement in scientific practices means not only attending to the product of their investigation—the scientific explanation—but also to the process by which it is constructed (Berland and Reiser 2009; Lehrer and Schauble 2006; Sandoval and Millwood 2005). This study illustrates how, given the opportunity to lead their own discussions within student-paced investigations, students can demonstrate sophisticated argumentation skills, even in the face of visually ambiguous information.

Of the skills apparent in Tad and Kino's discourse, one that is lacking is the ability to see another's perspective. Had the students truly considered the other's point of view, they may have come to see that both arguments were valid. However, that Tad and Kino were unsuccessful in persuading one another is less an indication of their ability to handle ambiguity, and more an indication of the limitations of the curriculum design. The unit exemplifies many other classroom-based curricula in its focus on capturing students' consensus. Whereas it prompted students to explain the process of their reasoning, it offered no alternative than for students to ultimately come to agreement. Given no room within the unit to record their dissent, nor the choice to distinguish their individual responses, Tad and Kino decided to take turns answering each question. That is, Tad selected his interpretation of changing global temperatures for the first question, and Kino specified his choice of evidence for the third question. To a researcher or teacher who had not witnessed their preceding discussion, this mismatch between claim and evidence would appear incoherent, rather than as a compromise resulting from a rich, but unresolved argument.

Long-standing scientific controversies are not unusual (e.g., how to measure quantum particles, what caused the extinction of the dinosaurs, whether or not parallel universes exist), and competing theories can co-exist for decades. Ambiguity in visual representations is common in professional science, and science instruction might be designed to acknowledge it so that it is not frustrating when students encounter it, too. Analysis of these students' discussion raises several issues for designing learning environments with ambiguous representations. For example, might the unit have been clearer in presenting the graph? For example, could the question have been better phrased, or the graph better contextualized, to help students avoid frustration over its ambiguity? On the other hand, might the unit have instead acknowledged and legitimized ambiguous interpretations, and offered space for students to explicitly document their individual ideas?

Regardless of how the unit might have been improved, the student-paced format of the GCC unit offered some benefit for supporting argumentation over traditional, teacher-led formats. A shortcoming of the initiate-respond-evaluate (IRE) pattern of whole-class instruction (Mehan 1979) is that students lose valuable opportunities to practice publicly defending and persuading others of their ideas (Berland and Reiser 2009). When pressed to cover large amounts of content material in little time, teachers often resort to immediately resolving students' conflicting ideas by giving the "correct" answer. It is rarer in such teacher-led discussions that students can pursue extended lines of argumentation (Radinsky et al. 2010). By not being prompted to reconcile various competing ideas, students rarely develop their points of view sufficiently to recognize their strengths and weaknesses, which research indicates is a valuable learning opportunity in disagreements (Bell and Linn 2000; de Vries et al. 2002; Scardamalia and Bereiter 1994). Without disagreement, students can miss chances to articulate connections between evidence and claims, which may otherwise lead to deep learning over shallow memorization of facts (Chi et al. 1994; Coleman 1998; Wells and Arauz 2006), and a shared understanding of the concepts being explored (Chin and Osborne 2010).

In contrast, computer-supported instruction affords many opportunities for students to build consensus that are not always possible during teacher-led classroom discussion. With technology to guide progress through an inquiry activity, students working at their own pace can engage in extended conversations, challenge one another's ideas, and seek new ways to articulate connections between evidence and claims. Ambiguity can spur this process of argumentation, and provide a context in which students become learning resources for one another (Dillenbourg et al. 1995).

Acknowledgements This research was supported by the National Science Foundation, grant number 0918743. A preliminary version of this work was presented at CSCL 2011, the Conference on Computer Supported Collaborative Learning.

Funding information Matuk, C. F., Sato, E., & Linn, M. C. (2011). *Agreeing to disagree: Challenges with ambiguity in visual evidence.* Proceedings of the 9th International conference on computer supported collaborative learning CSCL2011: Connecting computer supported collaborative learning to policy and practice, (Vol. 2, pp. 994–995). Hong Kong: The University of Hong Kong.

References

Amann, K., & Knorr Cetina, K. (1988). The fixation of (visual) evidence. *Human Studies, 11*(2), 133–169.

Avola, D., Caschera, M.C., Ferri, F., Grifoni, P. (2007). Ambiguities in sketch-based interfaces, Proceedings of the 40th Hawaii International Conference on System Science (HICSS '07), Hawaii.

Barthes, R. (1977). Rhetoric of the image. In S. Heath (Ed.), *Image, Music, Text.* New York: Hill and Wang.

Bell, P., & Linn, M. C. (2000). Scientific arguments as learning artifacts: Designing for learning from the web with KIE. *International Journal of Science Education, 22,* 797–817.

Berland, L. K., & Reiser, B. J. (2009). Making sense of argumentation and explanation. *Science Education, 93*(1), 26–55.

Bowen, G. M., Roth, W. M., & McGinn, M. K. (1999). Interpretations of graphs by university biology students and practicing scientists: Toward a social practice view of scientific representation practices. *Journal of Research in Science Teaching, 36*(9), 1020–1043.

Carter, B. (2006, April 9). There IS a problem with global warming... it stopped in 1998. *The Telegraph Newspaper.*

Chi, M. T. H., Leeuw, N. D., Chiu, M. H., & Lavancher, C. (1994). Eliciting self-explanations improves understanding. *Cognitive Science, 18*(3), 439–477.

Chin, C., & Osborne, J. (2010). Students' questions and discursive interaction: Their impact on argumentation during collaborative group discussions in science. *Journal of Research in Science Teaching, 47*(7), 883–908.

Coleman, E. B. (1998). Using explanatory knowledge during collaborative problem solving in science. *Journal of the Learning Sciences, 7*(3&4), 387–427.

Collins, H. M. (1998). The meaning of data: Open and closed evidential cultures in the search for gravitational waves. *American Journal of Sociology, 104*(2), 293–338.

de Vries, E., Lund, K., & Michael, B. (2002). Computer-mediated epistemic dialogue: Explanation and argumentation as vehicles for understanding scientific notions. *Journal of the Learning Sciences, 11*(1), 63–103.

Dillenbourg, P., Baker, M., Blaye, A., & O'Malley, C. (1995). The evolution of research on collaborative learning. In P. Reimann & H. Spada (Eds.), *Learning in humans and machines: Towards an interdisciplinary learning science* (pp. 189–211). Oxford: Elsevier.

Driver, R., Newton, P., & Osborne, J. (2000). Establishing the norms of scientific argumentation in classrooms. *Science Education, 84*(3), 287–312.

d'Ulizia, A., Grifoni, P., & Rafanelli, M. (2008). Visual notation interpretation and ambiguities. In F. Ferri (Ed.), *Visual languages for interactive computing: Definitions and formalizations. Information Science Reference* Hershey: IGI GLobal.

Easterling, D. R., & Wehner, M. F. (2009). Is the climate warming or cooling? *Geophysical Research Letters, 36,* L08706. https://doi.org/10.1029/2009GL037810.

Edwards, J. L., & Winkler, C. K. (1997). Representative form and the visual ideograph: The Iwo Jima image in editorial cartoons. *Quarterly Journal of Speech, 83*(3), 289–310.

Eisenberg, E. M. (1984). Ambiguity as strategy in organizational communication. *Communication monographs, 51*(3), 227–242.

Empson, W. (1932). *Seven types of ambiguity*. Cambridge: Cambridge University Press.

Eppler, M. J., Mengis, J., & Bresciani, S. (2008, July). Seven types of visual ambiguity: On the merits and risks of multiple interpretations of collaborative visualizations. In *Information Visualisation, 2008. IV'08. 12th International Conference* (pp. 391–396). IEEE.

Eppler, M. J., & Sukowski, O. (2000). Managing team knowledge: Core processes, tools and enabling factors. *European Management Journal, 18*(3), 334–342.

Fischer, F., Bruhn, J., Gräsel, C., & Mandl, H. (2002). Fostering collaborative knowledge construction with visualization tools. *Learning and Instruction, 12*(2), 213–232.

Friel, S. N., Curcio, F. R., & Bright, G. W. (2001). Making sense of graphs: Critical factors influencing comprehension and instructional implications. *Journal for Research in Mathematics Education, 32*, 124–158.

Futrelle, R.P. (2000). Ambiguity in visual language theory and its role in diagram parsing, IEEE Symposium on Visual Language, 172–175, Tokio IEEE Computer Society.

Gaver, W. W., Beaver, J., & Benford, S. (2003). Ambiguity as a resource for design, proceedings of the conference of human factors in computing system, 5–10 April 2003, Fort Lauderdale, FL. New York ACM Press.

Glazer, N. (2011). Challenges with graph interpretation: A review of the literature. *Studies in Science Education, 47*(2), 183–210. https://doi.org/10.1080/03057267.2011.605307.

Grunbaum, A. (1960). The Duhemian argument. *Philosophy of Science, 27*(1), 75–87.

Kaput, J. J. (1987). Representation and mathematics. In C. Janvier (Ed.), *Problems of representation in mathematics learning and problem solving* (pp. 19–26). Hillsdale: Erlbaum.

Karl, T. R., Arguez, A., Huang, B., Lawrimore, J. H., McMahon, J. R., Menne, M. J., et al. (2015). Possible artifacts of data biases in the recent global surface warming hiatus. *Science, 348*(6242), 1469–1472.

Kosslyn, S. M. (1989). Understanding charts and graphs. *Applied Cognitive Psychology, 3*(3), 185–225.

Latour, B., & Woolgar, S. (2013). *Laboratory life: The construction of scientific facts*. Princeton: Princeton University Press.

Lehrer, R., & Schauble, L. (2006). *Cultivating model-based reasoning in science education*. New York: Cambridge University Press.

Lewandowsky, S., & Spence, I. (1989). The perception of statistical graphs. *Sociological Methods & Research, 18*(2–3), 200–242. Chicago.

Linn, M. C., Eylon, B.–. S., & Davis, E. A. (2004). The knowledge integration perspective on learning. In M. C. Linn, E. A. Davis, & P. Bell (Eds.), *Internet environments for science education* (pp. 29–46). Mahwah: Erlbaum.

Mayer, R. E. (1993). Comprehension of graphics in texts: An overview. *Learning and Instruction, 3*, 239–245.

McNeill, K. L., & Krajcik, J. (2007). Middle school students' use of appropriate and inappropriate evidence in writing scientific explanations. In M. C. Lovett & P. Shah (Eds.), *Thinking with data: The proceedings of the 33rd Carnegie symposium on cognition* (pp. 233–265). Mahwah: Erlbaum.

McNeill, K. L., Lizotte, D. J., Krajcik, J., & Marx, R. W. (2006). Supporting students' construction of scientific explanations by fading scaffolds in instructional materials. *Journal of the Learning Sciences, 15*(2), 153–191.

Mehan, H. (1979). What time is it, Denise?: Asking known information questions in classroom discourse. *Theory Into Practice, 18*(4), 285–294.

Mooney, C. (2013, 7 October). Who created the global warming "pause"?. Mother Jones. Retrieved 27 July 2015 from http://www.motherjones.com/environment/2013/09/global-warming-pause-ipcc.

Mulkay, M. (1979). *Science and the sociology of knowledge*. London: George Allen and Unwin.

Nachmias, R., & Linn, M. C. (1987). Evaluations of science laboratory data: The role of computer-presented information. *Journal of Research in Science Teaching, 24*, 491–505.

National Research Council. (1996). *National Science Education Standards*. Washington, DC: The National Academies Press.

Nemirovsky, R., & Noble, T. (1997). On mathematical visualization and the place where we live. *Educational Studies in Mathematics, 33*(2), 99–131.

Nussbaum, E. M. (2008). Collaborative discourse, argumentation, and learning: Preface and literature review. *Contemporary Educational Psychology, 33*(3), 345–359.

Quintana, C., Eng, J., Carra, A., Wu, H., & Soloway, E. (1999). *Symphony: A case study in extending learner-centered design through process space analysis, paper presented at CHI 99: Conference on human factors in computing systems, may 19–21, 1999*. Pennsylvania: Pittsburgh.

Radinsky, J., Oliva, S., & Alamar, K. (2010). Camila, the earth, and the sun: Constructing an idea as shared intellectual property. *Journal of Research in Science Teaching, 47*(6), 619–642. https://doi.org/10.1002/tea.20354.

Reiser, B. J., Tabak, I., Sandoval, W. A., Smith, B., Steinmuller, F., & Leone, T. J. (2001). BGuILE: Stategic and conceptual scaffolds for scientific inquiry in biology classrooms. In S. M. Carver & D. Klahr (Eds.), *Cognition and instruction: Twenty five years of progress*. Mahvah: Erlbaum.

Robison, W., Boisjoly, R., & Hoeker, D. (2002). Representation and misrepresentation: Tufte and the Morton Thiokol engineers on the challenger. *Science and Engineering Ethics, 8*(1), 59–81.

Rye, J. A., Rubba, P. A., & Wiesenmayer, R. L. (1997). An investigation of middle school students' alternative conceptions of global warming. *International Journal of Science Education, 19*(5), 527–551.

Scardamalia, M., & Bereiter, C. (1994). Computer support for knowledge-building communities. *Journal of the Learning Sciences, 3*, 265–283.

Scardamalia, N., Bereiter, C., & Lamon, M. (1994). The CSILE Project: Trying to bring the classroom into the world. In K. McGilly (Ed.), *Classroom Lessons: Integrating Cognitive Theory and Classroom Practice*. Cambridge, MA: MIT Press.

Schoenfeld, A. H., Smith, J. P., & Arcavi, A. (1991). Learning: The microgenetic analysis of one student's evolving understanding of a complex subject matter domain. In R. Glaser (Ed.), *Advances in instructional psychology* (pp. 55–175). Hillsdale: Erlbaum.

Shah, P., Freedman, E. G., & Vekiri, I. (2005). The comprehension of quantitative information in graphical displays. In P. Shah & A. Miyake (Eds.), *The Cambridge handbook of visuospatial thinking* (pp. 426–476). New York: Cambridge University Press.

Shah, P., & Hoeffner, J. (2002). Review of graph comprehension research: Implications for instruction. *Educational Psychology Review, 14*(1), 47–69.

Sandoval, W. A., & Millwood, K. A. (2005). The quality of students' use of evidence in written scientific explanations. *Cognition and Instruction, 23*(1), 23–55.

Shepardson, D. P., Niyogi, D., Choi, S., & Charusombat, U. (2009). Seventh grade students' conceptions of global warming and climate change. *Environmental Education Research, 15*(5), 549–570.

Stocker, T. F., Qin, D., Plattner, G. K., Alexander, L. V., Allen, S. K., Bindoff, N. L., et al. (2013). Technical summary. In *Climate Change 2013: The Physical Science Basis. Contribution of Working Group I to the Fifth Assessment Report of the Intergovernmental Panel on Climate Change* (pp. 33–115). Cambridge University Press.

Svihla, V., & Linn, M. C. (2012). A design-based approach to fostering understanding of global climate change. *International Journal of Science Education, 34*(5), 651–676.

Tufte, E. (1997). *Visual explanations: Images and quantities, evidence and narrative*. Cheshire (CT): Graphics Press.

Tversky, B. (2002). Some ways that graphics communicate. In N. Allen (Ed.), *Working with words and images: New steps in an old dance*. Westport: Ablex Publishing Corporation.

Wells, G., & Mejía-Arauz, R. (2006). Toward dialogue in the classroom. *The Journal of the Learning Sciences, 15*(3), 379–428.

Winn, W. D. (1987). Charts, graphs and diagrams in educational materials. In D. M. Willows & H. A. Houghton (Eds.), *The psychology of illustration* (Vol. 1, pp. 152–198). New York: Springer.

Zacks, J., Levy, E., Tversky, B., & Schiano, D. (2002). *Graphs in print. In Diagrammatic representation and reasoning* (pp. 187–206). London: Springer.

A Learning Performance Perspective on Representational Competence in Elementary Science Education

Laura Zangori

For curriculum and instruction to foreground scientific literacy in elementary instruction, students require opportunities to develop understanding about how scientific domains and practices are interrelated and interconnected (American Association for the Advancement of Science [AAAS], 1993; National Research Council 2012). Yet within the elementary classroom, domain knowledge is presented in discrete pieces focusing on serration, classification, and observation that stop prior to engaging students in the knowledge-building practices of science, such as modeling and scientific explanations (Metz 2006, 2008). In this manner, science becomes known to elementary students as discrete pieces that "work" to produce outcomes without opportunities to engage in sense-making through using their evidence to understand *how* and *why* these pieces work. This results in fragmented knowledge rather than a robust foundation that serves to anchor their future learning in upper level course work (Duschl et al. 2007).

Elementary science learning environments were designed in this manner due to previous assumptions about students' sense-making abilities. Yet these assumptions were made on what early learners could do individually and did not consider what they can do when they are supported within their learning experience through curriculum, instruction, and their peers. As Duschl et al. (2007) states: "What children are capable of at a particular age is the result of a complex interplay between maturation, experience, and instruction" (p. 2). Research within elementary science learning environments indicates early learners develop epistemic understanding and engage in sense-making practices when provided space, opportunity, and support within their science learning environments (Lehrer and Schauble 2010; Manz 2012; Metz 2008; Ryu and Sandoval 2012; Zangori and Forbes 2014, 2016).

L. Zangori (✉)
University of Missouri, Columbia, MO, USA
e-mail: zangoril@missouri.edu

© Springer International Publishing AG, part of Springer Nature 2018
K. L. Daniel (ed.), *Towards a Framework for Representational Competence in Science Education*, Models and Modeling in Science Education 11,
https://doi.org/10.1007/978-3-319-89945-9_5

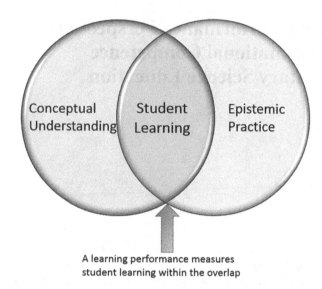

A learning performance measures
student learning within the overlap

Fig. 1 Learning performances measure student discipline-specific content situated in epistemic practice

A key sense-making practice is scientific modeling in which students construct models of their own design to use as explanatory tools. Their sense-making is situated within the explanatory power of the models they use to articulate connections between *what* occurred (observation and/investigation) and *how* and *why* it occurred (scientific explanation). This is their *model-based explanation* (Forbes et al. 2015; Schwarz et al. 2009; Zangori and Forbes 2016). As their knowledge about the discipline-specific content grows through collecting evidence and identifying causal mechanisms, they evaluate and revise their model to refine its use for sense-making (Clement 2000; Forbes et al. 2015; Gilbert 2004; Halloun 2007; Louca and Zacharia 2012; Schwarz et al. 2009).

The purpose of this chapter is to explore possible learning paths students may take as they understand the explanatory power of their models about the plant life cycle through the development and use of a *learning performance* (Krajcik et al. 2007) theoretical construct. Learning performances identify pedagogical maps as students are supported through curriculum and instruction to build conceptual understanding and reasoning about discipline-specific content (Krajcik et al. 2007). In essence, they are grade- and curriculum-focused learning progressions that are practice-based and grounded in evidence. In this manner, learning is situated within a discipline-specific epistemic practice to explore how elementary students engage in modeling, for example, to reason scientifically as shown in Fig. 1. The learning performance then serves to develop and/or modify the curriculum and instruction for the individual grade band. I will examine how elementary students engage with the epistemic practice of modeling, the conceptual understandings that elementary students bring with them into their modeling experiences, how to overlay these two constructs to build a learning performance, and how this information may be used to inform curriculum and instruction.

Epistemic Practice of Modeling

Model-based explanation construction about scientific phenomena is an essential activity in scientific learning environments (Gilbert et al. 2000). Expressed models are the physical manifestation of conceptual understanding and reasoning that the individual has internalized and holds about the scientific phenomenon (Halloun 2007; Louca and Zacharia 2012). The knowledge and reasoning in the expressed model is continually challenged on both the individual and social levels as new understandings are identified, evaluated, and revised (Nersessian 2002). Modeling artifacts are therefore historical records of both individual and social conceptual understanding and reasoning about scientific phenomena (Halloun 2007; Nersessian 2002).

Within science learning environments, this continual evaluation and revision of models is identified as the iterative nature of modeling. Students construct or develop an initial model in response to a question or problem about a scientific phenomenon. They use their model as a sense-making tool to understand how and why the phenomenon worked as it did. They then evaluate their model after completion of activities that provide them with new understanding. Finally they revise their model to reflect their new understanding and again use their model to articulate their new understanding (Clement 2000; Halloun 2007; Schwarz et al. 2009). Throughout each opportunity with modeling, students' epistemic and conceptual knowledge builds as they use their models as reasoning tools and evaluate if their models support their understanding in "why the phenomenon behaves as it does" (Gilbert et al. 2000, p. 196).

Putting Modeling into Practice

As shown in Fig. 2, I operationalize the epistemic facets of modeling through *developing, using, evaluating* and *revising* models (Clement 2000; Louca and Zacharia 2012; Schwarz et al. 2009). As students *develop, evaluate*, and *revise*, models are *used* for their explanatory power in which the model serves as a tool for sense-making. This occurs as students use their models to connect observation with theory to understand how and why the phenomenon occurred. It is the critical mode for model engagement because it is where students are using their representation to generate a model-based explanation (Schwarz et al. 2009).

First, learners *develop* their initial model in response to a question or problem. This initial model is the conceptual window into the learners' prior knowledge and reasoning about how and why the phenomenon behaves. This model is used to

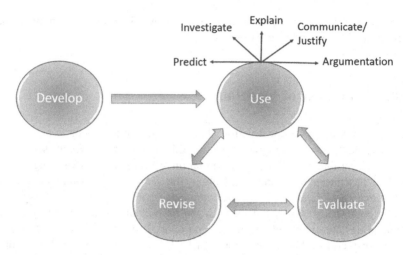

Fig. 2 A model of modeling practices for model-based explanations

generate predictions of how and why the phenomenon behaves. As new knowledge becomes available through investigations or observations, the initial constructed model is *evaluated* for the explanatory power it expresses. Based on the evaluation, the model may then be *revised* due to new understandings and ideas formed from the investigation and/or observations (Duschl et al. 2007). Within the classroom, each step supports students in making their thinking and reasoning visible. Learning occurs through the dialogic relationship between the student and the model as the student continually evaluates and revises the model based on their new knowledge (Louca and Zacharia 2012).

Modeling theory argues that engagement in scientific explanations through the practices of modeling *is* the very essence of scientific reasoning and learning (Gilbert 2004; Halloun 2007; Louca and Zacharia 2012). Constructed models are the physical expression of conceptual knowledge and scientific reasoning. Predictions are models built on prior or current knowledge, investigations are models-in-use under contrived conditions, outcomes are data models, and scientific explanations are new models extended from data models and based on new knowledge (Halloun 2007). Model-based learning occurs iteratively through revisiting and examining modeled constructions for adequacy to address the problem on both individual and social levels (Gilbert 2004). Conceptual understanding about the phenomenon is revised as new mechanisms become understood and accepted (Nersessian 2002). Within this framework, the epistemic practice of modeling theory equates to learning theory, and modeling artifacts are historical records of both individual and social conceptual understanding at different moments in time (Halloun 2007; Nersessian 2002).

Generating scientific explanations from constructed models is not intuitive; the ability to engage in scientific reasoning requires both knowledge of the domain and knowledge of the practice (Nersessian 2002). In the absence of either, scientific modeling is then reduced to illustrations, demonstrations, or summaries that are not sufficient for the learner to *use* to articulate how and why a process occurs. This is frequently what occurs in the elementary classroom where fostering and promoting scientific reasoning and scientific modeling are rare (Lehrer and Schauble 2010; Schwarz et al. 2009). For example, process illustrations of the water cycle and plant growth are abundant within elementary science learning environments, yet these diagrams are used to identify what students should "know" about the process through iconic illustrations. These models are frequently used in the classroom as memorization devices. The water cycle becomes reduced to students identifying condensation, evaporation, and precipitation without opportunities to connect *what* is occurring with *how* and *why* it is occurring. The plant cycle is reduced to stages that appear in discrete bursts; life cycle diagrams begin with a seed and the next picture is a sprout and then the final picture is a full-grown plant. Yet these diagrams do not provide opportunities for students to reason about spatial or temporal boundaries of growth and development. Without curriculum and instruction that includes opportunities to develop, use, evaluate, and revise their own models for knowledge building, elementary students lack opportunities to build a foundation for model-based reasoning.

While there are few studies involving early learners in the practices of modeling, empirical findings suggest that when early elementary students are first asked to develop 2-D or 3-D models to determine how and why a scientific phenomenon occurs, they attempt to recreate what the phenomenon looks like rather than identify how and why it works (Lehrer and Schauble 2010; Schwarz et al. 2009). However, through successive opportunities with modeling and support through curriculum and instruction, elementary students develop an understanding that their attempts to mimic what they see does not provide how and why processes function. They then begin to revise their models to build their own understanding (Lehrer and Schauble 2010; Manz 2012). Further, as this epistemic understanding about the practices of modeling grows, so does conceptual understanding about scientific phenomena, and vice-versa (Manz 2012; Ryu and Sandoval 2012).

Learning Performances and Learning Progressions

Learning performances are situated within the learning progression construct, but they are not the same construct. While learning progressions are a macro-level attempt to map conceptual development of content and practices across broad time spans, learning performances provide a micro-level focus on crossing a single big idea with a single scientific practice within a single grade band (Duncan et al. 2009; Krajcik et al. 2007). For example, a learning performance may encompass 3rd-grade students' abilities to construct and use models to scientifically reason (i.e., scientific practice) about plant growth and development (i.e., big idea) over the course of an instructional unit. This learning performance may eventually serve as an anchor on a learning progression on student reasoning about ecosystems.

However, learning progressions measure snapshots in time across multiple grade-bands about what students know and how they reason with their knowledge. The difference is that learning performances are not *what students know* but the paths they take in *coming to know*. These paths can then be used to define curriculum and instruction for the grade level to identify where students' knowledge requires bolstering or where it may be leveraged to support their understanding of new knowledge. Learning performances are maps of children's possible conceptual framework development as their understanding about particular concepts and practices grows over the curriculum (Alonzo and Steedle 2009; Duschl et al. 2007). The pathways students take through the performance will not be identical, as learners' conceptual understanding varies depending on their prior beliefs, knowledge and existing conceptual frameworks (Duschl et al. 2007). However, attempting to establish identical paths is not the goal. Rather, the overall goal is to empirically examine the most likely path students may encounter *en route* to building a conceptual framework so that they are scaffolded in building a successively more mature conceptual understanding over time (Shin et al. 2010).

Content-focused learning progressions show how students use model-based reasoning to understand different science concepts across wide time-spans. This work finds that students rarely attain model-based reasoning. Learning progressions that began in 4th grade (Mohan et al. 2009), 5th grade (Duncan et al. 2009; Gunckel et al. 2012), and 7th grade (Stevens et al. 2010) through upper secondary school and undergraduate students (i.e., Stevens et al. 2010) have examined if and how students use model-based reasoning to understand different science concepts. The learning progressions show that models and modeling are not considered as sense-making tools by elementary, middle, or high school students. The engagement in modeling was often similar across the grade-levels, even though content knowledge was increasing. These learning progressions identify that curriculum and instructional guidance is required to support students in building their knowledge within the practices of modeling.

The results of these learning progressions are not surprising. Across all grade levels, teachers struggle to understand the purpose and utility of models and modeling (Justi and Gilbert 2002; Oh and Oh 2011). When standard curriculum includes models, it is rare that it includes opportunities for students to gather information about the model or produce their own understanding using the model (Duschl et al. 2007; Gilbert 2004). The nature and purpose of modeling are not typically included in curriculum and instruction and students are not typically asked to test the explanatory power of their models (Louca and Zacharia 2012; Schwarz et al. 2009). Furthermore, what students are able to do at each grade level is critically dependent on the learning environment (Duschl et al. 2007; Gilbert 2004; Louca and Zacharia 2012; Manz 2012). All of these things contribute to student learning, yet these are not specifically addressed within learning progressions. Learning progressions tell us where we should be. The goal of the learning performance is to identify where students are at the beginning of the instruction, where they arrive at the end of instruction, and how curriculum and instruction supports them on their path to the upper anchor of a learning progression.

Building a Learning Performance

To develop a learning performance (e.g., Forbes et al. 2015; Zangori & Forbes 2016), we use _construct centered design,_ or CCD (Shin et al. 2010). CCD provides an empirically-tested framework that situates the development of the learning performance within science content and scientific practice. In this framework, first the construct is selected (the "big" idea) and defined, then the appropriate content and practice standards are identified and unpacked to identify target explanations (Krajcik et al. 2007). The target explanations embedded within the scientific practice are used to design a hypothetical learning performance for the possible learning paths students might take to reach the target explanation. Next, a student task is implemented in the classroom to examine student learning of the construct through scientific practice. Students' ideas about the construct are then used to ground the learning performance; in response, the learning performance is revised in an iterative cycle as empirical data become available and the learning performance is refined. Finally, the learning performance and all associated materials are reviewed by an external source. The stages of learning performance development are operationalized through four steps.

Step 1: Select and Define Construct

To demonstrate how conceptual understanding develops within the practices of modeling, I use the big idea of plant growth and development. The _Next Generation Science Standards_ (NGSS Lead States 2013) state that elementary students in grades 3 through 5 should develop models about organism life cycles. For plant life, this includes the seed, adult plant, seed production, seed dispersal, death of adult, and seedling growth. These standards, as well as those from the _Atlas for Scientific Literacy_ (AAAS 2007), were identified and unpacked (See Table 1).

Table 1 Unpacking the relevant standards

Standard core idea	Standard reference	Scientific practice	Science literacy map	Unpacking statements
LS1.B: Growth and development of organisms	3-LS1–1	Developing and Using Models	The living environment, flow of matter in ecosystems map	Use models to understand and reason that plants undergo a predictable life cycle that includes birth (seed germination), development, and death. Through fruiting and see dispersal, offspring grow and the life cycle returns to a starting state so the species lives but the individual plant may die.

Step 2: Create Claims through Development of a Hypothetical Learning Performance

Targets for the ways in which students engage in the modeling practices to build understanding about the plant life cycle were then defined from the unpacked standards so that a tentative hypothetical learning performance framework could be proposed. This initial attempt at the learning performance is a standard component in learning performance development as it provides the starting point for examining student understanding about this big idea through epistemic practice. The development of the hypothetical learning performance here is grounded in the literature on elementary students' engagement in scientific reasoning (Duschl et al. 2007; Metz 2008; Manz 2012), elementary student learning (Duschl et al. 2007; Lehrer and Schauble 2010; NGSS Lead States 2013), modeling (Gilbert 2004; Clement 2000; Schwarz et al. 2009), and conceptual understanding of plant growth and development (Manz 2012; Metz 2008; NGSS Lead States 2013; Zangori and Forbes 2014). The hypothetical framework is presented in Table 2.

Step 3: Specify Evidence and Define Tasks

Since learning performances are a measure of *knowledge-in-use* serving to identify how students build and use knowledge of discipline-specific concepts through epistemic activity (Krajcik et al. 2007), the student tasks required that students engage in the modeling practices to understand the plant life cycle. A widely available set of curriculum materials, *Structure of Life* ([SOL], FOSS 2009), was chosen for their

Table 2 Learning performances framework for 3rd grade model-based explanations about plant concepts

Modeling feature	Plant life cycle
Develop	Develop models to understand how and why plants grow and change in predictable ways through a life cycle that includes birth (seed germination), development, and death.
Use	Use models to reason how and why plants grow and change in predictable ways through a life cycle that includes birth (seed germination), development, and death.
Evaluate	Evaluate models for how well they support understanding and reasoning about how and why plants grow and change in predictable ways through a life cycle that includes birth (seed germination), development, and death.
Revise	Revise models to better support understanding and reasoning about how and why plants grow and change in predictable ways through a life cycle that includes birth (seed germination), development, and death.

lessons. However, this curricular unit does not provide explicit opportunities to engage in the practices of modeling and scientific explanations (FOSS 2009; Metz 2006). To make these practices explicit, we highlighted each practice during the lessons and included how the practice would help students understand the plant life cycle. In addition, three supplemental model-based explanation lessons (SML) were embedded within the curricular unit. The SMLs were enacted by the classroom teachers at three time points over the course of their SOL curriculum enactments: immediately after introducing the students to the curriculum ideas in Investigation 1, after completion of Investigation 1, and immediately following completion of Investigation 2. The SMLs were grounded in research on modeling practices in the elementary classroom (Forbes et al. 2015; Lehrer and Schauble 2010; Schwarz et al. 2009). They were situated within the SOL curriculum unit and aligned with the FOSS lesson structure.

To support the teachers in using the SMLs, all teachers in the study participated in a professional development workshop on modeling in the summer prior to data collection. The modeling experiences from the professional development were frequently discussed with the teachers and, throughout the study, the author worked one-on-one with each teacher to support their practice of the supplemental lessons and model-based teaching and learning. The teachers were provided the SML materials at least 2 months in advance and, even though they all had prior experience with similar lessons (e.g., Forbes et al. 2015), the SML materials were reviewed with each teacher prior to each enactment. The lessons provided background information on the practices of scientific modeling (construct, use, evaluate, and revise) and instructions specific to creating 2-D diagrammatic process models.

Each SML began with the teachers holding whole class discussions to elicit students' ideas about what might be a model. Prompts were included to facilitate a student discussion about modeling prior to students beginning the modeling task. The student discussion prompts began with eliciting students' ideas about models by asking students to consider examples which included:

- observing a bird's behavior at a bird feeder,
- drawing a food chain,
- going on a field trip to the Grand Canyon,
- making a bridge out of toothpicks and testing how much weight it can hold, and
- doing an experiment to investigate growth.

After students considered each example, the teachers asked the students explain their thinking to determine how they decided whether or not something was a model. The discussion prompts for the whole class discussion included, "Which of these examples are models? Why? What is a model? What do models look like? and How do you think scientists use models?" The student ideas were listed on the SmartBoard or other classroom devices so that all students could see their ideas, allowing the teacher to lead a whole class discussion. At the end of each discussion, the teachers saved the discussion ideas so they could be revisited during the subsequent SMLs so students could discuss if and how their thinking was changing during the unit. At the end of the discussion, the teachers emphasized:

- A model makes really complicated things in nature simpler so we can understand and study them.
- Scientists construct, use, evaluate, and revise models to explain and make predictions about natural phenomena.
- People can create models with pictures, words, mathematical equations, and computer programs.
- We can use models in the classroom to help us understand seed growth.
- We can use our models to share ideas and to make those ideas better. We can get new or different ideas from other people, and we can think through our own ideas when people ask us about our models.

At the end of each SML, students were given packets to complete a modeling task involving constructing a 2-D diagrammatic process model using pencils in response to the question, "How does a seed grow?" Once their models were drawn, they wrote responses to a series of reflective questions designed to elicit the epistemic considerations comprising model-based explanations accompanying their models: (a) "What does your model show happening to a seed? (b) Why do you think this happens to a seed? (c) What have you seen that makes you think this is what happens to a seed? and (d) How would you use your model to explain to others how a seed grows?" In addition, prior to model 2 and 3, students were asked to return to their previous model and evaluate it for how they thought it showed what was happening to the seed and to write what changes they thought they needed to make in order to show where seeds come from.

Step 4: Empirical Grounding of the Learning Performance

To identify the learning paths that 3rd-grade students travel within the practices of modeling to build understanding about the plant life cycle, the learning performance was empirically grounded through students' discussions about their models and writings. Students were interviewed immediately following each modeling task. Through their discussions, three measurable levels for each modeling practice were identified. The levels were empirically grounded through an iterative process moving between the students' discussions and the theoretical learning performance to fine-tune the learning performance so that it captured the ways in which student engaged with the practices of modeling to learn about the plant life cycle. The learning performance were then submitted for external review. The final learning performance is presented in Table 3.

To understand how students come to understand modeling, two students, Daisy and Dahlia, are followed through their articulation of the learning performance levels across the three modeling iterations.

Table 3 Learning performance for 3rd-grade students' engagement in the modeling practices to learn about the plant life cycle

Modeling feature	Plant life cycle
Develop/ Revise	Level 1: Developing/revising models that are literal (concrete) illustrations of a plant.
	Level 2: Developing/revising models to replicate a model of the plant life cycle that includes birth (seed germination), development, and death.
	Level 3: Developing/revising models to understand abstractions of the plant life cycle that includes birth (seed germination), development, and death.
Use	Level 1: Using models to show a plant.
	Level 2: Using models to show what I know about the model of plant life cycle.
	Level 3: Using models to reason about how and why plants grow and change in predictable ways through a life cycle that includes birth (seed germination), development, and death.
	Level 1: Evaluating models for concrete illustration of a plant.
Evaluate	Level 2: Evaluating models for their ability to replicate a model of the plant life cycle.
	Level 3: Evaluating models for how well they support understanding and reasoning about how and why plants grow and change in predictable ways through a life cycle that includes birth (seed germination), development, and death.

Daisy

At the beginning of the study, Daisy exhibited a Level 1 understanding for the model feature, *Develop/Revise*. Her first model was of a single plant. Within this level, Daisy included sun, rain, and some indication of a seed, but she did not discuss these parts when talking about her model. Daisy's discussion was about that she developed a drawing to show a plant and did not include that she drew her picture for any purpose. Within her discussion about her model, she did not include any other items that were on her drawing. Overall, within this level, she identified that she was only showing a picture of a plant (Fig. 3).

During her discussion, Daisy only *used* her model to show the plant. For example, when asked to use her model to explain how a seed grows, she identified that her model showed "roots coming out and a plant after that". Even though Daisy's model also included the sun and rain which were important for plant growth, she did not use these items to explain their connection to the plant she drew or to make sense of how and why water and sunlight were necessary for the plant.

Before Daisy drew her second model, she went back to her first model to evaluate it for how well it showed where seeds come from. When asked about how she evaluated her first model, she focused on her concrete illustration of a plant, a Level 1 of *Evaluation*. She stated that her first model did not show enough "details" about the plant and in her second model she would "put more details" so the plants would look "more realistic." However, it is important to note that while her evaluation

Fig. 3 Daisy's model (W_10M) scoring at Level 1 for develop/revise models of the plant life cycle

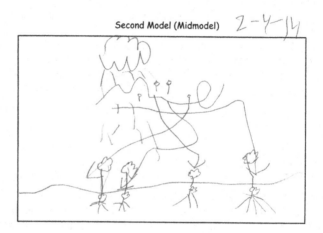

Fig. 4 Daisy's second model

criteria for her first model was at a Level 1 focusing on concrete details, her second model (Fig. 4) showed a shift in her ideas from a concrete illustration to a replicate model of plant growth. With this shift, Daisy also started to use her model to "show what she knows" and replicated seed growth. She stated in her interview about her second model "…it [the plant] has seeds in it like seed pods and it [the plant] grows and they'll fall out…depends on how windy it is how far they [seeds] go." She was replicating the process of seed dispersal and how seeds end up in different locations, as we also did in the SOL curriculum. The arrows on her model point down to the plant showing that each seed dispersing in the wind is connected to an adult plant, yet she did not include how this portion of the process occurred. Her model was attempting to replicate the process presented in the curriculum. Her second model is at a Level 2 for *Develop/Revise* because while it did show seed dispersal, she did

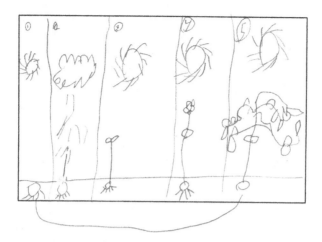

Fig. 5 Daisy's final model

not connect seed dispersal to how and why plants grow and change, only showing that there was a connection between seeds, wind, and adult plants.

By the end of the study, Daisy's discussion about her third model (Fig. 5) was at a Level 2 for all modeling features, *Develop/Revise, Use*, and *Evaluate*. She used all elements of her model stating, "It [the plant] gets sun and rain to grow leaves" as the plant grows from seed to adult. Her final model was a replication of plant growth, showing discrete bursts separated by lines. While the model included abstractions such as rain and sun, she did not consider these things for how or why they might explain change in the plant across the five panels, such as sunlight to make food or rain being absorbed by roots to help the plant grow. She also included a line moving from the final panel to the first panel indicating that plant life may be in a cycle, but this connection was not discussed by her nor did she identify where the initial seed comes from, even though her second model showed seed dispersal. When asked "Where do seeds come from?" she stated that they "come from people or they fly by themselves or something." While she has connected some knowledge between model 2 and 3, this connection was not on her model, instead drawing a replication of an iconic plant life cycle.

In *Using* her final model, Daisy is also a Level 2. She represented elements because she considered them important to her model, but she only used these elements to show what she knows, not to predict or explain how and why the elements supported the plant life cycle. For example, when asked if water was a necessary component for the plant life cycle, Daisy responded that she knew that plants need water and sunlight, but not how or why these things are important. When including in her discussion these items on her drawing, she identified that she included sun and rain because she *should* rather than reasoning about how they were necessary for plant growth.

Finally, Daisy's evaluation criteria for her second model was how well it mimicked replications of seed growth (a Level 2), and being sure that she added elements that better represent seed growth:

I: So is there much that has changed between your...models?
Daisy: This one [model one] doesn't have any buds. And a seed pod.
I: Why did you show buds and the seed pod in your second model?
Daisy: Because I learned most of this in class.

She also focused her evaluation on replicating what she learned in class rather than evaluating her model on how it helped her form understandings about the plant life cycle.

In summary, Daisy's learning path for *Developing/Revising, Using, and Evaluating* models began at Level 1 where she drew a single plant with abiotic elements (sun and water), but did not consider how or why these elements should be on her model. Daisy's second model also included an abiotic process, but shifted to a Level 2 and considered what was necessary for seeds to spread. Daisy's final model is also a Level 2, showing a complete plant life cycle, identifying that water and sunlight are involved in this process, and including numbered stages on her model indicating that plant growth takes time. However, while her final model was a complete life cycle, she had not *Developed, Evaluated*, and/or *Revised* the model to help her make sense of these things. Her modeling was to replicate what she knew and had seen in class rather than the model becoming a tool to represent the process that she could use for her own knowledge building. Even though her final model includes abiotic processes that should provide explanatory power to the model, Daisy added these elements (rain and sun) because she was supposed to rather than to help her understand the plant life cycle process.

Dahlia

All modeling features surrounding Dahlia's first model were at a Level 2 (Fig. 6). Dahlia's initial discussions identified that the nature and purpose of her models was to accurately represent an iconic representation of the plant life cycle showing stages of growth. As Dahlia stated in her discussion about her initial model, "So first it's a seed then um [*sic*] it starts growing a little stem and then stem and leaves and then it grows to a plant." As seen on Dahlia's model (Fig. 6), she has separated time across the seed growing to a fully-grown flower. During her interview, Dahlia identified that everything she chose to include in her model was information that she knew about plant growth. She used her model as a device to show what she knew about iconic seed growth rather than using her model as an explanatory device to understand how or why these things occur.

The bigger it gets a stage higher it gets

Fig. 6 Dahlia's first model (A_13M)

Dahlia *used* the detail she included in her model to also show what she knew about the plant life cycle. For example:

I: Tell me about your model.
Dahlia: I put...put detail into it.
I: Why was it important to put more detail in it?
Dahlia: Cause it helps me remember it.

Throughout Dahlia's discussions about her first model, she continually identified that it was important that she add detail so she could use her model if she forgot about how the plant life cycle occurs. She discussed that if she, or someone else, wanted to learn about how seeds grow, then they could look at her detailed illustration or read the descriptions she included on her pictures to learn about the plant life cycle.

Dahlia's evaluation of her first model focused on information she did not know in model one but did know for model two (Fig. 7). Her evaluation criteria also supported her ideas about replicating an iconic image of the plant life cycle rather than representing the plant life cycle such as:

I: What did you think of evaluating your model?
Dahlia: I liked it.
I: Why did you like it?
Dahlia: Because it [my model] was detailed.

Dahlia's evaluation of her first model was positive in that it closely matched a replication of the plant life cycle. Her ideas about her model as a memory device remained through all modeling features of her first model, including *Evaluating* her first model, and align with iconic models of plant growth which are frequently used by students as memory devices.

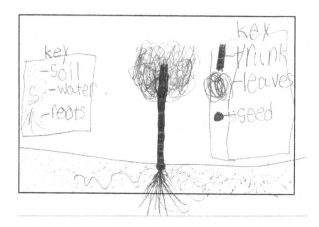

Fig. 7 Dahlia's second model

However, there is a shift in Dahlia's ideas about modeling when she finished her second model. Her focus became on developing a representation rather than a replication. This shift was evidenced in the elements she included in her second model and her discussion about her second model (Fig. 7). She included abstracted abiotic processes necessary for plant growth, such as water underground, and considers why including these processes was necessary to plant life. For example, when asked about the water she has included within the ground, she stated:

> Like [*sic*] the roots grow and then it [the roots] goes into the soil to hold it [the tree] down and then like the rain falls. It might not get exactly to the seed and it might go everywhere and then the roots will get it…the water from the ground so the seed can grow.

She was not showing abiotic processes because she is supposed to, rather she was *using* them to understand how these abiotic processes affect plant growth. While her second model did not show growth from seed to an adult, she considered necessary requirements for seed growth and using her model to articulate how seeds and plants get water.

When Dahlia discussed her final model, she talked about how she was attempting to make sense of what was occurring in the plant life cycle as it happens in the natural world. She discussed her model as a representation of the plant life cycle rather than as iconic replications. Dahlia's final model, shown in Fig. 8, still represented the iconic levels of plant growth from her first model, but she has now added rain and sun. However, when Dahlia talked about this final model, she identified that when she was drawing this model, she wanted the model to help her "understand that if it stays a seed forever, then the next cycle won't go on." This statement indicates Dahlia sees that her model has the ability to be a generalization for supporting her ideas about plant growth (all plants have a cycle) as well as identifying that her model is related to the physical world (there are different kinds of plants) (See Fig. 8).

Furthermore, the ways in which she discussed the purposes of developing and revising her model shifted to identifying that models hold both observable and non-observable elements:

Fig. 8 Dahlia's final model

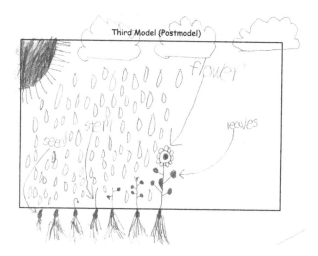

I: Tell me about your model.

Dahlia: Um...where should I start? Well first the seeds [*sic*] planted and then the seed coat breaks off so the roots can come off and then the roots suck up water so the embryo can grow and then the roots suck up more water so the embryo can grow out and then it gets more water so the leaves can grow out and the secondary stems. And then it keeps getting water and sunlight and then all of a sudden the flower and buds come out and the leaves get big enough so they can start making their own food and the flower and the leaves start getting more food and water and sun.

As Dahlia discussed her model, she included the non-observable (embryo, water, and sunlight) that have become visible to her. Dahlia also indicated she was using her model as a sense-making tool about the plant life cycle. Immediately following that discussion about her model revision, Dahlia also articulated how and why the plant cycle occurred: "Um [*sic*] it just keeps going and the plant cycle just like buds and seeds fall off and get planted again then they grow." This suggests she was using her representation to propose a scientific explanation for the plant life cycle, which she understood as that the cycle has to keep going or the plants will die. Dahlia's model was starting to become a tool to help her make sense of the plant life cycle.

Dahlia's ideas about *Developing* and *Using* a model also carried into her final model evaluation. After Dahlia finished discussing her third model, she was asked to go back to evaluate her second model and discuss why she made changes between the second and third model:

I: What changes did you make to your third model?

Dahlia: Germination...A seed ... Seed with stem ... Flower.

I: Okay. Why is it important to know germination?

Dahlia: So you know how trees, and flowers, Earth, and everything grows.

Furthermore, throughout the interview, she continued to *Evaluate* and *Revise* her model as she spoke about the plant life cycle. She pointed out things on her model that she should also include and change, such as showing seed pods and a new seed sprouting. At Level 3, the model became a dynamic representation for her, in which she continued to *Use* to make sense of the plant life cycle, and *evaluate* and *revise* as her thinking clarified throughout our discussion.

In summary, even though many of the elements Dahlia included on her model did not change from her first to final model, her ideas about what those elements meant to her *did* change. Dahlia started at a Level 2 for all modeling features, in which she Developed, Used, and Evaluated her model for how well it served as a mimic of an iconic model of plant growth. Yet her thinking about her model began to shift to a Level 3 in the second modeling iteration in which she began to consider abstract processes, such as water underground, and considered how these processes supported her thinking about plant growth. By the final model iteration, she was at a Level 3 for *Revise, Use*, and *Evaluate* as her model became dynamic, and she used her model throughout her final interview as a thinking tool to help her understand how and why plants grow.

Discussion

The completed learning performance provided a foundation on which to understand the ways in which students engage with the practices of modeling to conceptualize and reason about plant processes. The case studies of the two students, Daisy and Dahlia, served to examine the learning paths that these two 3rd-grade students traveled for *developing, using, evaluating*, and *revising* their models for sense-making over the course of an instructional unit about plant growth and development. Since the learning performance was grounded in the students' use of epistemic practice to learn about a core idea within their classroom context, it provided learning paths within classroom norms (Krajcik et al. 2007).

To separate content from process and only examine how students engage in a particular scientific practice or what students' ideas are about certain content does not give a complete picture of *what* students conceptual resources are or *how they come* to evaluate and build upon their pre-existing ideas (Duschl et al. 2007; Forbes et al. 2015; Metz 2008; Manz 2012). Therefore, learning performances also provide a means to examine students' pre-existing ideas and how these ideas are developed and refined through the practices of modeling. As such, learning performances anchor learning progressions for what might be assessed at the grade level, identify how to support students in meeting that assessment, and provide stepping stones for what students should be able to do at subsequent grade levels (Duncan et al. 2009).

In using the learning performance to examine how Daisy and Dahlia discussed their models, the case studies demonstrated their growth in how they came to understand the purpose of modeling as well as understanding the processes necessary for the plant life cycle. What is most striking is that the purpose both students understood for *developing* their model (concrete, iconic, or abstract) also defined how they *used* their model to understand and reason about what plants need to grow and survive, and what elements they chose to evaluate within their models. When Daisy saw her model as an iconic representation of the plant life cycle, then her model *use* became focused on how her model looked like, or replicated, a plant growth model. Once Dahlia identified her model as an abstraction, then her *use* and *evaluation* of her model included consideration of how she could use her model to help her make sense of the plant life cycle.

How to conceptually shift students from the notion of a model as a static detailed illustration to the notion that models are knowledge-building cultural tools of science is a persistent challenge within model-based teaching and learning (Justi and Gilbert 2002; Oh and Oh 2011; Nersessian 2002; Lehrer and Schauble 2010; Schwarz et al. 2009). The learning performance presented here suggests that these changes are small and gradual, but may occur over the course of a curricular unit. Students initially build their conceptions about what a model is from pictures and objects that are used to illustrate. As evidenced here, even though students' initial models held necessary elements for explanatory power, they did not identify them as such because they did not yet recognize that they could be used as such. However, over time, Dahlia's drawing began to stand for something and became dynamic. As she discussed her model, she grabbed a pencil or traced with her finger how things moved and grew on her model, made changes to her model as she spoke, and included "hidden" movement below ground or up in the sky in her model. She identified the things she had drawn were real occurrences but not identical to nature because she was showing more than she could observe outside.

This shift in the students thinking about the nature and purpose of modeling occurred with continued support from the classroom science learning environment. Scientific epistemic and conceptual knowledge is both individual and socially influenced by what students experience, understand and reason about on the social plane, and then internalize and form into discipline-specific epistemic knowledge they employ for their individual models (Halloun 2007; Louca and Zacharia 2012; Nersessian 2002). The curriculum included space and time for students to build understanding about modeling and to use their models during their investigations. The more opportunities the students were provided to use their models, the greater the impact on their understanding of the modeling practice and their model use for sense-making.

Models and modeling has come to the forefront as a key sense-making practice with the incorporation of scientific modeling as one of the NGSS essential practices. Yet, sense-making has long been absent in elementary science curriculum materials (Metz 2006, 2008; Forbes et al. 2015; Lehrer and Schauble 2010; Zangori and Forbes 2014) so these practices must be incorporated by teachers (e.g., Forbes et al. 2015). The incorporation of modeling experiences must be meaningful so that students have a purpose for creating the model rather than going through the motions (Lehrer and Schauble 2010; Schwarz et al. 2009). Models should be in multiple forms, such as concrete and abstract, but should also be made explicit within curriculum and instruction so that students began to understand the nature and purpose of constructing and revising models to provide explanatory power.

The lessons should provide students opportunities to build knowledge about multifaceted and complex systems so that both content and epistemic knowledge about modeling can grow together (Lehrer and Schauble 2010; Louca and Zacharia 2012; Manz 2012). Scientific epistemic and conceptual knowledge is both individual and socially influenced by what students experience, understand and reason about on the social plane, and then internalize and form into discipline-specific epistemic knowledge they employ for their individual models (Halloun 2007; Louca and Zacharia 2012; Nersessian 2002).

Conclusion

The field has much to learn about how to optimally support early learners to use the scientific practice of modeling to build scientifically acceptable conceptions of and to reason scientifically about biological systems and the interconnectedness with other systems. Learning progression research has identified the difficulties elementary, middle, secondary, and undergraduate students have regarding the nature and purpose of models. Learning performances are now required to identify how to support students in meeting the targets identified in the learning progressions. Learning performances not only serve to identify the learning paths of children for curriculum development and instructional support, but they also provide a scaffold for teacher knowledge and practice about the topic as well as a pedagogical map for ways to respond and scaffold student progress within the domain and scientific practice (Duncan et al. 2009; Duschl et al. 2007). This research provides important insights into future work to support teachers in helping them understand how to provide opportunities and promote students' sense-making about plant systems within the practices of modeling and where changes and scaffolds to curriculum and instruction might be most effective.

Acknowledgments This paper is based on the author's doctoral dissertation. This research was supported in part by the Paul and Edith Babson Fellowship and the Warren and Edith Day Doctoral Dissertation Travel Award, University of Nebraska-Lincoln. I appreciate the interest and participation of the students and teachers who made this work possible. I also thank Patricia Friedrichsen for insightful comments on an earlier draft of the manuscript.

References

Alonzo, A. C., & Steedle, J. T. (2009). Developing and assessing a force and motion learning progression. *Science Education, 93*(3), 389–421. https://doi.org/10.1002/sce.20303.

American Association for the Advancement of Science. (1993). *Benchmarks for science literacy*. New York, NY: Oxford University Press.

American Association for the Advancement of Science. (2007). Atlas for scientific literacy. Washington, D.C.

Clement, J. (2000). Model based learning as a key research area for science education. *International Journal of Science Education, 22*(9), 1041–1053. https://doi.org/10.1080/095006900416901.

Duncan, R. G., Rogat, A. D., & Yarden, A. (2009). A learning progression for deepening students understandings of modern genetics across the 5th-10th grades. *Journal of Research in Science Teaching, 46*(6), 655–674. https://doi.org/10.1002/tea.20312.

Duschl, R. A., Schweingruber, H. A., & Schouse, A. W. (2007). *Taking science to school: Learning and teaching science in grades K-8.* Washington, D.C: The National Academies Press.

Forbes, C. T., Zangori, L., & Schwarz, C. V. (2015). Empirical validation of integrated learning performances for hydrologic phenomena: 3rd-grade students' model-driven explanation-construction. *Journal of Research in Science Teaching, 52*(7), 895–921. https://doi.org/10.1002/tea.21226.

Full Option Science Systems. (2009). *Teacher guide: Structures of life.* Berkeley: Delta Education.

Gilbert, J. (2004). Models and modelling: Routes to more authentic science education. *International Journal of Science and Mathematics Education, 2*(2), 115–130. https://doi.org/10.1007/s10763-004-3186-4.

Gilbert, J., Boulter, C., & Rutherford, M. (2000). Explanations with models in science education. In J. K. Gilbert & C. J. Boulter (Eds.), *Developing models in science education* (pp. 193–208). Netherlands: Kluwer Academic Publishers.

Gunckel, K. L., Covitt, B. A., Salinas, I., & Anderson, C. W. (2012). A learning progression for water in socio-ecological systems. *Journal of Research in Science Teaching, 49*(9), 843–868. https://doi.org/10.1002/tea.21024.

Halloun, I. A. (2007). Mediated modeling in science education. *Science and Education, 16*(7–8), 653–697. https://doi.org/10.1007/s11191-006-9004-3.

Justi, R. S., & Gilbert, J. K. (2002). Modelling, teachers' views on the nature of modelling, and implications for the education of modellers. *International Journal of Science Education, 24*(4), 369–387. https://doi.org/10.1080/09500690110110142.

Krajcik, J., McNeill, K. L., & Reiser, B. J. (2007). Learning-goals-driven design model: Developing curriculum materials that align with national standards and incorporate project-based pedagogy. *Science Education, 92*(1), 1–32. https://doi.org/10.1002/sce.20240.

Lead States, N. G. S. S. (2013). *Next generation science standards: For states, by States.* Washington, DC: Electronic Book, National Academies Press.

Lehrer, R., & Schauble, L. (2010). What kind of explanation is a model? In M. K. Stein & L. Kucan (Eds.), *Instructional explanations in the disciplines* (pp. 9–22). Boston: Springer US.

Louca, L. T., & Zacharia, Z. C. (2012). Modeling-based learning in science education: Cognitive, metacognitive, social, material and epistemological contributions. *Educational Review, 64*(4), 471–492.

Manz, E. (2012). Understanding the codevelopment of modeling practice and ecological knowledge. *Science Education, 96*(6), 1071–1105. https://doi.org/10.1002/sce.21030.

Metz, K. E. (2006). The knowledge building enterprises in science and elementary school science classrooms. In L. B. Flick & N. G. Lederman (Eds.), *Scientific inquiry and nature of science* (pp. 105–130). Netherlands: Springer.

Metz, K. E. (2008). Narrowing the gulf between the practices of science and the elementary school science classroom. *The Elementary School Journal, 109*(2), 138–161.

Mohan, L., Chen, J., & Anderson, C. W. (2009). Developing a multi-year learning progression for carbon cycling in socio-ecological systems. *Journal of Research in Science Teaching, 46*(6), 675–698. https://doi.org/10.1002/tea.20314.

National Research Council. (2012). A framework for K-12 science education: Practices; cross-cutting concepts; and core ideas. Washington, D.C.

Nersessian, N. J. (2002). The cognitive basis of model-based reasoning in science. In P. Carruthers, S. Stich, & M. Siegal (Eds.), *The cognitive basis of science* (pp. 133–153). Cambridge, UK: Cambridge University Press.

Oh, P. S., & Oh, S. J. (2011). What teachers of science need to know about models: An overview. *International Journal of Science Education, 33*(8), 1109–1130. https://doi.org/10.1080/09500693.2010.502191.

Ryu, S., & Sandoval, W. A. (2012). Improvements to elementary children's epistemic understanding from sustained argumentation. *Science Education, 96*(3), 488–526. https://doi.org/10.1002/sce.21006.

Schwarz, C. V., Reiser, B. J., Davis, E. A., Kenyon, L., Acher, A., Fortus, D., & Krajcik, J. (2009). Developing a learning progression for scientific modeling: Making scientific modeling accessible and meaningful for learners. *Journal of Research in Science Teaching, 46*(6), 632–654. https://doi.org/10.1002/tea.20311.

Shin, N., Stevens, S. Y., & Krajcik, J. (2010). Tracking student learning over time using construct-centered design. In S. Rodrigues (Ed.), *Using analytical frameworks for classroom research. Collecting data analysing narrative* (pp. 38–58). New York: Routledge.

Stevens, S. Y., Delgado, C., & Krajcik, J. S. (2010). Developing a hypothetical multi-dimensional learning progression for the nature of matter. *Journal of Research in Science Teaching, 47*(6), 687–715. https://doi.org/10.1002/tea.20324.

Zangori, L., & Forbes, C. T. (2014). Scientific practices in elementary classrooms: Third-grade students' scientific explanations for seed structure and function. *Science Education, 98*(4), 614–639.

Zangori, L., & Forbes, C. T. (2016). Development of an empirically based learning performances framework for 3rd-grade students' model-based explanations about plant processes. *Science Education, 100*(6), 961–982. https://doi.org/10.1002/sce.21238.

Part II
Teaching Towards Representational Competence

Supporting Representational Competences Through Adaptive Educational Technologies

Martina A. Rau

Connection Making Between Multiple Graphical Representations: A Multi-Methods Approach for Domain-Specific Grounding of an Intelligent Tutoring System for Chemistry

Introduction

External representations are ubiquitous in science, technology, engineering, and mathematics (STEM) domains. Instructional materials in these domains rely on external representations to illustrate abstract concepts and mechanisms that constitute the domain-relevant content knowledge (Arcavi 2003; Cook et al. 2007; Kordaki 2010; Lewalter 2003). Typically, instructors present students with not only one representation but with multiple because different representations provide complementary information about the to-be-learned concepts (Kozma et al. 2000; Larkin and Simon 1987; Schnotz and Bannert 2003; Zhang 1997; Zhang and Norman 1994). For example, when students learn about chemical bonding, they typically encounter the representations in Fig. 1 (Kozma et al. 2000): Lewis structures and ball-and-stick figures show bond types, ball-and-stick figures and space-filing models show the geometrical arrangement of the atoms, and electrostatic potential maps (EPMs) use color to show how electrons are distributed in the molecule. When students learn about fractions, they typically use the representations shown in Fig. 2 (Cramer 2001; Siegler et al. 2010): circles diagrams depict fractions as equally sized parts of an inherent "whole" (i.e., the shape of a full circle), rectangle diagrams show

M. A. Rau (✉)
University of Wisconsin, Madison, Madison, WI, USA
e-mail: marau@wisc.edu

© Springer International Publishing AG, part of Springer Nature 2018   103
K. L. Daniel (ed.), *Towards a Framework for Representational Competence in Science Education*, Models and Modeling in Science Education 11,
https://doi.org/10.1007/978-3-319-89945-9_6

Fig. 1 Representations of ethyne: Lewis structure, ball-and-stick figure, space-filling model, electrostatic potential map (EPM)

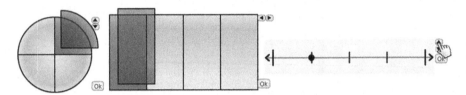

Fig. 2 Representations of fractions: circle, rectangle, and number line

fractions as parts of a continuous whole that can be divided along two dimensions, and number lines show fractions as measures of proportions where the unit is defined as a standard length. Indeed, there is an immense literature documenting the potential benefits of multiple representations on students' learning (Ainsworth 2006; de Jong et al. 1998; Eilam and Poyas 2008).

However, students' benefit from multiple representations depends on their ability to integrate the different conceptual perspectives into a coherent mental model of the domain knowledge (Ainsworth 2006; Gilbert 2008; Schnotz 2005). To do so, students need to make connections between representations (Ainsworth 2006; Bodemer and Faust 2006; Someren et al. 1998; Wu et al. 2001). Connection making is a representational competence that students struggle to attain, even though it is critical to their learning (Ainsworth et al. 2002; Rau et al. 2014a, b). Students' low connection-making competences are considered a main obstacle to success in STEM domains because they jeopardize students' learning of important domain concepts (Dori and Barak 2001; Moss 2005; Taber 2001; Talanquer 2013).

Hence, students need to receive instructional support to acquire connection-making competences (Ainsworth 2006; Bodemer and Faust 2006; Vreman-de Olde and Jong 2007). Prior research in many domains shows that students' learning of domain knowledge can be enhanced by providing instructional support for connection-making competences (Berthold et al. 2008; Linenberger and Bretz 2012; Rau et al. 2012; Seufert and Brünken 2006; van der Meij and de Jong 2006).

Recent research indicates that adaptive educational technologies such as intelligent tutoring systems (ITSs) can be particularly effective in supporting connection-making competences (Rau et al. 2015; Rau et al. 2012). ITSs support step-by-step problem solving (VanLehn 2011) and provide adaptive instructional support (Corbett et al. 2001; Koedinger and Corbett 2006). Adaptive support in ITSs typically includes feedback upon the diagnosis of a student's misconception (e.g., based on certain errors he/she makes while solving a problem), hints on demand (e.g., the student requests help on solving a step), and problem selection (e.g., based on the student's diagnosed knowledge level, the tutor selects a new problem that is considered to be of appropriate difficulty). These adaptive capabilities contribute to the effectiveness of ITSs, which have been shown to significantly enhance learning in a variety of STEM domains (VanLehn 2011).

A particular strength of adaptive educational technologies in supporting students' connection-making competences is that they can model and assess students' ongoing acquisition of these competences while they engage in learning activities that target domain-relevant concepts. Based on these ongoing assessments, they can adapt the type of support students receive in real time. The main promise of such adaptive technologies is based on the idea that—because connection.

making competences are an important aspect of domain learning—helping students to learn these competences will enhance their learning of the domain knowledge. In designing adaptive support for connection-making competences, we need to address the following three questions:

1. Which connection-making competences do we need to support?
2. How should we design activities in educational technologies so that they effectively support students' acquisition of these competences?
3. What is the relationship between these competences and other student characteristics such as mental rotation ability and prior domain knowledge? The remainder of this chapter is structured along these three guiding questions. I conclude the chapter by discussing preliminary principles for the design of adaptive support for connection-making competences and directions for future research.

Which Connection-Making Competences Do We Need to Support?

When deciding which connection-making competences we should support, it is helpful to first consider what characterizes expertise in STEM domains. Since the main purpose in using multiple representations is to enhance learning of domain knowledge, expertise in connection making is inevitably tied to domain expertise: learning to make connections between domain-specific representations results in learning of the domain-relevant concepts that the representations depict. In defining expertise in connection-making competences, I draw on Ainsworth's (2006) Design, Functions, and Tasks (DeFT) framework, on Kellman and colleagues perceptual learning paradigm (Kellman and Garrigan 2009; Kellman and Massey 2013), on science and math education research that focuses on representational competences (Cramer 2001; Kozma and Russell 2005; Pape and Tchoshanov 2001; Patel and Dexter 2014), and on research on domain expertise (Dreyfus and Dreyfus 1986; Gibson 1969, 2000; Richman et al. 1996).

One aspect of expertise is *conceptual understanding* of connections (Ainsworth 2006; Kozma and Russell 2005; Patel and Dexter 2014). An expert conceptually understands connections based on relations between features of different representations that show corresponding concepts. For example, a chemistry expert who encounters the representations shown in Fig. 1 conceptually understands which perceptual features of different representations convey corresponding information about domain-relevant concepts (i.e., information that is shown by both representations; e.g., hydrogen is shown as Hs in Lewis structures and as white spheres in ball-and-stick figures), and which features convey complementary information about domain-relevant concepts (i.e., information that is shown by one representation but not by the other; space-filling models do not differentiate between bond types, but Lewis structures and ball-and-stick figures do). This conceptual understanding of connections is part of the expert's mental model of domain-relevant concepts (e.g., the expert integrated information about bond type from the Lewis structure with information about electron density distributions from the EPM into their conceptual understanding of why compounds with triple bonds tend to be highly reactive).

Consider another example in which a math expert is presented with a number line that shows a dot at 1/4 and a circle diagram that is 1/4 shaded (see Fig. 2). The expert conceptually understands why both representations show the same fraction by connecting the shaded section in the circle to the section between zero and the dot in the number line, because both features depict the numerator. Further, the expert relates the number of total sections in the circle to the sections between 0 and 1 in the number line, because both features show the denominator. The expert's ability to make these connections exhibit conceptual understanding of domain-relevant concepts; namely, that a fraction is a portion of something relative to something else.

The importance of conceptual understanding of connections between representations is widely recognized as an important aspect of domain expertise in the literature on learning with external representations (Ainsworth 2006; Patel and Dexter 2014), in the literature on expertise (Dreyfus and Dreyfus 1986; Richman et al. 1996), in science education (Jones et al. 2005; Wu and Shah 2004), and in math education (Charalambous and Pitta-Pantazi 2007; Cramer 2001). Furthermore, educational practice guides emphasize the importance of helping students conceptually understand connections between representations (e.g., National Council of Teachers of Mathematics; NCTM 2000, 2006).

A second aspect of expertise is *perceptual fluency* in translating between representations. Perceptual learning processes first received attention by Gibson (1969), who investigated how visual perception improves with experience. The fact that experts perceive information differently than novice students is well researched (e.g., Chi et al. 1981; Gegenfurtner et al. 2011). Perceptual learning paradigms are founded on the observation that domain experts "see at a glance" what a given representation shows. This high degree of fluency in processing visual representations has been defined as the improved ability to recognize visual patterns that results from extensive practice (Gibson 1969). With respect to connection making, perceptual fluency allows students to automatically see whether two representations show the same information and to combine information from two representations efficiently without any perceived mental effort. For example, a chemistry expert who sees the Lewis structure and EPM of ethylene shown in Fig. 1 will automatically identify the triple bond in the Lewis structure and the region of high electron density shown in red in the EPM as relevant features. By relating these features to one another, the expert automatically see that the triple bond in the Lewis structure and the red region in the EPM show the same thing; namely, the fact that triple bonds result in high electron density in that region, which results in attraction of less electronegative compounds and results in reactions with these compounds. An expert engages in these connection-making processes quickly and automatically, without perceived mental effort. Consider again the math expert who encounters the representations shown in Fig. 2. The expert automatically sees that the circle and the number line show the same fraction, without having to engage in reasoning about how they cover the same proportion of area or of length. These two examples illustrate that perceptual connection-making competences involve (1) recognizing which feature of a given representation depicts domain-relevant information, (2) mapping this feature to the corresponding feature in another representation, (3) integrating information from different representations into mental models of domain-relevant information, and (4) fluency in these perceptual processes (Kellman and Massey 2013). Thus, by engaging in perceptual learning processes in connection making, students acquire the ability to engage in perceptual processes automatically, quickly, with ease, and with little mental effort. Perceptual learning processes result in perceptual fluency in making connections that frees up cognitive capacities that experts can devote to complex conceptual reasoning about the domain knowledge (Kellman and Massey 2013). The importance of perceptual fluency in making connections for domain expertise is widely recognized in research on expertise (Dreyfus and Dreyfus 1986; Gibson 1969, 2000; Richman et al. 1996), in science education (Kozma and Russell 2005; Wu and Shah 2004), and in math education (Pape and Tchoshanov 2001).

In summary, both conceptual understanding and perceptual fluency are important connection-making competences.

How should We Design Support for Connection-Making Competences? Two Intelligent Tutoring Systems

In this section, I review research on how we should design instructional support that is effective in promoting students' acquisition of conceptual connection-making competences and perceptual connection-making competences. In doing so, I will present two example ITSs that provide support for conceptual and perceptual connection-making competences.

Support for Conceptual Connection-Making Competences

Instructional support for conceptual processes helps students relate representations based on features that show corresponding domain-relevant concepts. Such conceptual support is effective in promoting students' learning of the domain knowledge (Bodemer, Ploetzner, Feuerlein, & Spada, Bodemer et al. 2004; Seufert and Brünken 2006; Van Labeke and Ainsworth 2002; Vreman-de Olde and Jong 2007). Educational technologies offer effective ways to support conceptual connection-making competences because they can employ dynamic color highlighting (Mayer 2003), dynamic linking (Ainsworth 2008a, 2008c; Bodemer et al. 2004; de Jong and van Joolingen 1998; van der Meij and de Jong 2006), and animations (Ainsworth 2008b; Betrancourt 2005; Holzinger et al. 2008).

Prior research on conceptual connection-making support yields several instructional design principles. First, conceptual support typically asks students to explain what features of representations depict *corresponding concepts* by mapping structurally relevant features (Gentner 1983; Seufert 2003; Seufert and Brünken 2006). Second, conceptual support seems to be particularly effective if it provides *prompts to self-explain* the mappings between representations. For example, Berthold and Renkl (2009) show that self-explanation prompts increase students' benefit from multiple representations. Self-explanation prompts are more effective if they ask students to self-explain specific connections than if they are open-ended (Berthold et al. 2008; Van der Meij & de Jong, 2011). Third, conceptual support should provide students with *assistance* in making connections because students typically struggle in making connections (Ainsworth et al. 2002), especially if they have low prior content knowledge (Stern et al. 2003). In line with these findings, research shows that assistance is particularly important for students with low prior content knowledge (Bodemer and Faust 2006), and when problems are particularly complex (van der Meij and de Jong 2006).

Open questions about how best to design conceptual support regard how much exploration students should be allowed to engage in versus how much structure students should receive. On the one hand, exploratory types of conceptual support provide little structure for students' interactions as they make connections. One common type of exploratory conceptual support is auto-linked representations. Auto-linked representations are dynamically linked such that the student's manipulations of one representation are automatically reflected in the other representation (e.g., Van Labeke and Ainsworth 2002; van der Meij and de Jong 2006, 2011). Thus, students can explore how intermediate steps, mistakes, and the final result look like in two or more linked representations. On the other hand, structured types of conceptual support allow for little exploration as students make connections. Structured types of conceptual support typically use un-linked representations but provide step-by-step guidance to make sense of corresponding elements shown in the different representations (e.g., Bodemer and Faust 2006; Bodemer et al. 2004; Bodemer et al. 2005; Gutwill et al. 1999; Özgün-Koca 2008). This question of how to balance structure and exploration has been discussed under the term "assistance dilemma"

(Koedinger and Aleven 2007). On the one hand, too much structure and too little exploration may fail to engage students in actively processing the instructional materials (Koedinger and Aleven 2007). On the other hand, too little structure and too much exploration can lead to cognitive overload for students, jeopardizing their learning gains (Schwonke et al. 2011). It is possible that a student's prior knowledge influences whether he/she will benefit from more exploration or more structure. For example, a student who has prior experience with the individual representations may require less structure and may benefit from exploring correspondences because he/she understand each representation by itself well. By contrast, a student who is making connections between representations when one or more of them are relatively novel might need more structure to succeed at this task. However, this question has not yet been resolved in research on conceptual support for connection making.

I now turn to discussing two example ITSs that implement conceptual connection-making support: one ITS for elementary-school fractions learning and one ITS for college-level chemistry learning.

The Fractions Tutor

The Fractions Tutor is an effective ITS designed for use in real classrooms with elementary-school students (Rau et al. 2012; Rau et al. 2013). It covers a range of topics typically covered in elementary school curricula, ranging from naming fractions to fraction subtraction. It uses commonly used graphical representations of fractions (e.g., see Fig. 2). The Fractions Tutor is available for free at www. fractions.cs.cmu.edu. Students log onto the website with personal logins and work on the tutor problems individually at their own pace. Teachers are provided with a tool that allows them to retrieve information about their students' performance, for instance, how many errors students made on a particular set of problems. They can use this information to identify what problem type a particular student struggles with, and provide targeted advice to that individual student. The Fractions Tutor supports learning through problem solving while providing just-in-time feedback and on-demand hints. In contrast to other ITSs, which often support procedural learning, the Fractions Tutor emphasizes conceptual learning by focusing on conceptual interpretations of fractions as proportions of a unit while students solve problems.

The Fractions Tutor features problems designed to support conceptual connection-making competences. Figure 3 shows an example of a tutor problem in which students learn about the inverse relationship between the magnitude of the fraction and the denominator. Students are first presented with a worked example that uses one of the area models (i.e., circle or rectangle) to demonstrate how to solve a fractions problem. Students complete the last step of the worked example problem (with the worked example still on the screen) and are then presented with an isomorphic problem in which they have to use the number-line. Students have to

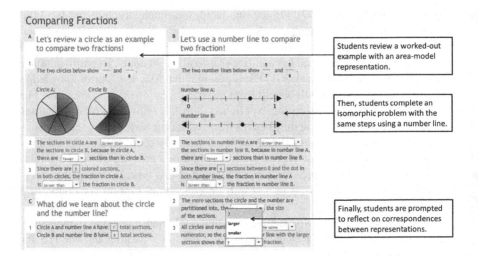

Fig. 3 Example conceptual connection-making problem: students construct a representation of a fraction

complete this isomorphic problem themselves. At the end of each problem students receive prompts to self-explain how to relate the two representations to one another by reasoning about how they depict fractions. The prompts in these problems use a fill-in-the-gap format with menu-based selection to support self-explanation.

The design of these problems aligns with the prior research on conceptual support just reviewed. To help students map *corresponding concepts* (e.g., that a larger denominator results in smaller sections), the problem-solving steps with the two representations are directly aligned. Furthermore, the *self-explanation prompts* are designed to help students relate the specific features of each representation to the abstract concepts targeted in the problem (e.g., the inverse relationship between the magnitude of the fraction and the denominator). Students receive *assistance* from the system in making connections: the Fractions Tutor provides detailed hints on demand for each step, and error feedback messages that tailor to the specific mistake the student made.

The Fraction Tutor's conceptual connection-making problems provide relatively much structure for students' reasoning about connections and leave relatively little room for exploration. This choice is based on empirical research that compared the effectiveness of conceptual problems with step-by-step guidance for mapping un-linked representations to conceptual problems that allowed students to explore mappings with auto-linked representations (Rau et al. 2012). The results from this study showed that for a sample of 4th- and 5th-grade students, step-by-step guidance with un-linked representations was more effective than exploration with auto-linked representations.

Chem Tutor

Chem Tutor is an effective ITS for undergraduate chemistry learning (Rau et al. 2015). The design of Chem Tutor is based on surveys of undergraduate chemistry students, interviews and eye-tracking studies with undergraduate and graduate students, and extensive pilot testing in the lab and in the field (Rau and Evenstone 2014; Rau et al. 2015). Chem Tutor covers topics of atomic structure and bonding. It features graphical representations typically used in instructional materials for these topics, including the representations shown in Fig. 1. It is available for free at https://chem.tutorshop.web.cmu.edu/. Chem Tutor has been used for homework assignments in introductory chemistry courses at the college level. Students can log into the system with their personal user accounts and complete problem sets that their instructor assigned to them. It is also possible to integrate Chem Tutor with learning management platforms such as Moodle (https://moodle.org/).

Chem Tutor features problems designed to support conceptual connection-making competences. These problems are designed to help students map features of different representations that show corresponding concepts. Figure 4 shows two example problems in which students construct a representation of an atom using an interactive tool, based on a given representation.

The design of these problems aligns with the prior research on conceptual support just reviewed. To help students map *corresponding concepts*, students are presented with two representations side-by-side. They have to use a given representation to construct a second representation. The *self-explanation prompts* were designed based on interviews with chemistry experts (Rau and Evenstone 2014; Rau et al. 2015), which showed that two types of reasoning play a role in conceptual connection-making competences. First, knowing about interchangeable functions of representations is important: students need to know which representations provide the same information about a concept (e.g., both the Lewis structure and the ball-and-stick figure show atom identity). Second, knowing about complementary functions of representations is important: students need to know which representations provide different information about a concept (e.g., the space-filling model

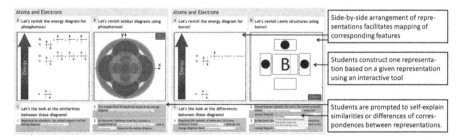

Fig. 4 Example conceptual connection-making problem: students construct one representation based on a given representation and self-explain similarities (left) or differences between representations

shows atomic volume but the Lewis structure does not). Thus, some of Chem Tutor's conceptual problems prompt students to self-explain similarities between representations, whereas other problems prompt students to self-explain differences between representations. Finally, Chem Tutor provides *assistance* in the form of hints and error feedback that target a misconception the system diagnosed based on the student's problem-solving behavior.

Support for Perceptual Connection-Making Competences

A separate line of prior research has focused on perceptual connection-making processes (Kellman and Garrigan 2009; Kellman and Massey 2013; Wise et al. 2000). In spite of the importance of perceptual processes for expertise (e.g., Even 1998; Kozma and Russell 2005; Pape and Tchoshanov 2001; Wu et al. 2001), instruction in STEM domains typically focuses on conceptual rather than on perceptual connection-making processes (Kellman and Garrigan 2009). Perhaps as a result from this focus, even proficient students tend to perform poorly on perceptual fluency tests, and their performance on perceptual fluency tests remains low throughout formal education (Kellman et al. 2009). To address this gap, Kellman and colleagues initiated a research program that aims at integrating support for connection making that is specifically tailored to perceptual processes into instruction for complex STEM domains (Kellman and Massey 2013). The main hypothesis of this research program is that perceptual support for connection making can significantly enhance educational outcomes in STEM domains.

To investigate this hypothesis, Kellman and colleagues developed interventions that provide perceptual support for a variety of mathematics and science topics (Kellman et al. 2009; Kellman et al. 2008; Wise et al. 2000). These interventions ask students to rapidly classify representations over many short trials while providing correctness feedback. Such trials are designed to expose students to systematic variation, often in the form of contrasting cases, so that irrelevant features vary but relevant features appear across several trials (Massey et al. 2011). Integrating perceptual support into educational technologies can be particularly effective because they can adapt to the individual student's learning rate. Indeed, in research with 6th- and 7th-graders, Massey et al. (2011) found that there was large variance in the number of practice problems students needed to solve to achieve a high level of perceptual fluency.

Kellman and colleagues conducted a number of controlled experiments and observational studies to investigate the effectiveness of perceptual support in in several domains, including mathematics (Kellman et al. 2008, 2009; Wise et al. 2000) and chemistry (Wise et al. 2000). Results from these studies showed that perceptual support for connection making leads to large and lasting increases in perceptual fluency (Kellman et al. 2008, 2009; Wise et al. 2000). More importantly, perhaps, gains in perceptual fluency were found to transfer to new examples that students did not encounter during training (Kellman et al. 2008) and, in some cases, lead to better performance in solving domain-relevant problems (Kellman et al. 2008; Wise et al. 2000).

I now describe how the Fractions Tutor and Chem Tutor implement perceptual connection-making support.

The Fractions Tutor

The Fractions Tutor features problems that are designed to help students become more fluent and efficient at translating between representations. In line with Kellman et al.' (2009) perceptual learning paradigm, the perceptual connection-making problems provide students with numerous practice opportunities to identifying corresponding representations. Hence, in our study, students receive numerous short classification problems. Figure 5 shows an example of a perceptual problem for the equivalent fractions topic of the tutor. Students sort a variety of representations into bins that show the same proportion. Students receive only correctness feedback, but no principle-based guidance for solving the problems, so as to encourage perceptual problem-solving strategies. Students can request hints, but hint messages only provide general encouragement (e.g., "give it a try!"). Finally, the perceptual learning paradigm emphasizes the importance of directing students' attention to perceptually relevant features. Thus, the perceptual problems encourage students to employ visual rather than conceptual strategies. For example, students are asked to visually judge equivalence rather than counting sections. To discourage counting strategies, the perceptual problems include representations with sections too small to count.

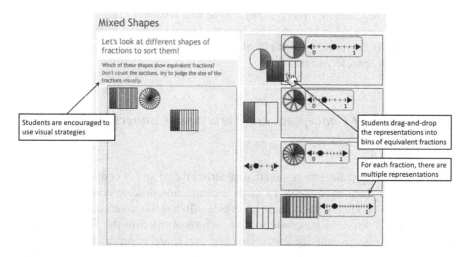

Fig. 5 Example perceptual connection-making problem: students use a drag-and-drop tool to sort representations into bins that show equivalent fractions

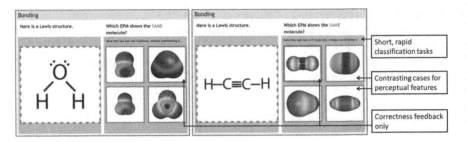

Fig. 6 Example perceptual connection-making problem: students select which of four alternative representations shows the same molecule as a given representation

Chem Tutor

Chem Tutor also provides problems that support students in acquiring perceptual connection-making competences. Chem Tutor provides students with numerous short classification problems. Consider the two problems shown in Fig. 6. Given one representation (e.g., a Lewis structure), students have to identify which of four other representations shows the same molecule (EPMs). Before each problem starts, students see a screen with a prompt to "solve this problem fast, without overthinking it", to encourage them to rely on perceptual cues to solve the problem.

The design of the perceptual problems is based on Kellman and colleagues perceptual learning paradigm (Kellman and Garrigan 2009; Kellman and Massey 2013; Massey et al. 2011). In particular, the different choice options provide variations of irrelevant features of the representations and contrasted perceptual features that provide relevant information (e.g., geometry, location of the local charges). Each problem is short (i.e., it involves only one step). Students receive several of these problems in a row, and they receive only correctness feedback. Thus, the perceptual problems are designed to help students become faster and more efficient at extracting relevant information from graphical representations based on repeated experience with a large variety of problems.

Effectiveness of Conceptual and Perceptual Connection-Making Support

As mentioned, the main assumption that underlies the design of connection-making support is that helping students to acquire connection-making competences will enhance their learning of the domain knowledge. To test this assumption, an experiment with 428 4th- and 5th-grade students who worked with the Fractions Tutor evaluated the effectiveness of conceptual and perceptual connection-making support on students' learning of fractions knowledge (see Rau et al. 2012 for a full report on this experiment). Students were drawn from five elementary schools in

Pennsylvania, USA. In the given academic year, the 4th- and 5th-grade students in this school district's scores on the mathematics Pennsylvania State Standardized Test were ranked as below basic for 0.8%, basic for 4.7%, proficient for 22.4%, and advanced for 72.1%. Students worked with the Fractions Tutor for 10 h during their regular math instruction, spread across consecutive days. Prior to the study, students took a pretest that assessed their knowledge about fractions concepts and fractions procedures. After they finished their work with the Fractions Tutor, they took isomorphic posttests. One week later, they took isomorphic delayed posttests. Students were randomly assigned to a different version of the Fractions Tutor that (1) did not include connection-making problems, (2) included only conceptual connection-making problems, (3) included only perceptual connection-making problems, or (4) included both conceptual and perceptual connection-making problems. In addition to the connection-making problems, all students received regular tutor problems that included only one representation per problem (i.e., either a circle diagram, rectangle diagram, or number line). Students in the control condition without connection-making support worked only on regular tutor problems. Students in all conditions spent the same amount of time working with the Fractions Tutor. The results showed a significant interaction effect between conceptual and perceptual connection-making support ($p < 0.05$), such that students who received both conceptual and perceptual connection-making support showed the highest learning gains. On fractions concepts, this group showed an improvement of 34% on the immediate posttest and of 48% on the delayed posttest, relative to their pretest performance. On fractions procedures, this group showed an improvement of 22% on the immediate posttest and of 28% on the delayed posttest, relative to their pretest performance. Thus, the results from this experiment are in line with the assumption that conceptual and perceptual connection-making support enhances students' learning of the domain knowledge.

An experiment with Chem Tutor evaluated the effectiveness of conceptual and perceptual connection-making support on students' learning of chemistry concepts (see Rau and Wu 2015 for a full report on this experiment). A total of 117 undergraduate students participated in the experiment in a laboratory. Students were recruited with posters and by advertising in introductory chemistry courses. 79% of the students were currently enrolled in general chemistry for non-science majors, 13.4% were enrolled in general chemistry for science majors, 2.5% were enrolled in advanced general chemistry, and 5% were not currently enrolled in a chemistry course. They worked on Chem Tutor's atoms and electrons unit for 3 h, spread across two sessions that were scheduled no more than 3 days apart. In the first session, students first took a pretest about chemistry concepts. They then worked through half of the Chem Tutor problems and took an intermediate posttest. When they came back for the second session, they worked through the remainder of the tutor problems and took a final posttest. Students were randomly assigned to a different version of Chem Tutor that (1) did not include connection-making problems, (2) included only conceptual connection-making problems, (3) included only perceptual connection-making problems, or (4) included both conceptual and perceptual connection-making problems. In addition to the connection-making problems, all students received regular tutor problems that included only one representation per problem. Students in the control condition without connection-making support worked only on regular tutor problems. All students spent the same amount of time working with Chem Tutor. The results replicated the findings with the Fractions Tutor: there was a significant interaction effect between conceptual and perceptual connection-making support ($p < 0.05$), such that students who received both conceptual and perceptual connection-making support showed the highest learning gains. This group showed an improvement of 45% on the final posttest, relative to their pretest performance. Thus, the results from this experiment are also in line with the assumption that conceptual and perceptual connection-making support enhances students' learning of the domain knowledge.

Taken together, two experiments with different student populations.

(elementary-school students and university undergraduate students), different domains (fractions and chemistry), and different settings (classroom and laboratory) provide evidence that helping students acquire conceptual and perceptual connection-making competences improves their learning of the domain knowledge.

What is the Relationship between these Competences and Other Student Characteristics?

Before we can design adaptive support for conceptual and perceptual connection-making competences, we need to know how they interact with one another, how the effectiveness of such support might depend on other student characteristics such as mental rotation ability and prior domain knowledge. In this section, I summarize a number of experiments that have investigated these relationships.

The Role of Mental Rotation Ability

Mental rotation ability is known to predict learning success in STEM domains (Uttal et al. 2013; Wai et al. 2009). The impact of mental rotation ability on students' learning of the domain knowledge may be of particular importance when students make connections between multiple graphical representations because this task requires students to make sense of the visuo-spatial relationships depicted in the different graphical representations (Stieff 2007). Integrating information across different graphical representations into a mental model of the domain knowledge requires students to map relevant features across representations. To do so, students need to hold the relative location of the features depicted in working memory and to mentally rotate these features so that they can be mapped to one another (Hegarty and Waller 2005). The cognitive load imposed by this task is arguably higher for students with low mental rotation ability than for students with high mental rotation

Fig. 7 Effect of perceptual support by mental rotation ability (0-33th, 34th–66th, and 67th–100th percentiles) on learning gains on the final posttest. Error bars show standard errors of the mean

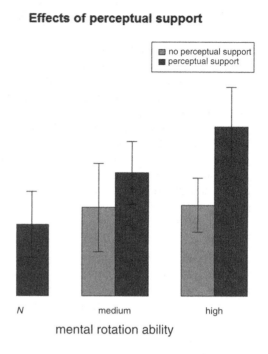

Effects of perceptual support

□ no perceptual support
■ perceptual support

N   medium   high

mental rotation ability

ability (Uttal et al. 2013). As a consequence, students with low mental rotation ability may fail at this task, which might jeopardize their learning success (Hegarty and Waller 2005; Stieff 2007; Uttal et al. 2013). Therefore, it is important to investigate whether students' mental rotation ability affects their benefit from conceptual and perceptual connection-making support.

The evaluation experiment with Chem Tutor mentioned above (see section 0; Rau and Wu 2015) investigated this question. Students in this experiment took the Vandenberg & Kuse mental rotation ability test (Peters et al. 1995) before they started working with Chem Tutor. The results from this experiment revealed no significant interactions between conceptual connection-making support and mental rotation ability ($p > 0.10$). However, there was a significant interaction of perceptual connection-making support with mental rotation ability ($p < 0.01$): perceptual connection-making support was effective for students with high mental rotation ability, but not for students with low mental rotation ability. Figure 7 illustrates the nature of this interaction effect.

This finding suggests that particularly perceptual connection-making problems are difficult for students with low mental rotation ability. In these problems, students have to quickly find matching representations, and these representations are not always spatially aligned, so that students have to mentally rotate the representations to find the matching one. This task may be more difficult for these students with low mental rotation ability, so they are more likely to fail at this task, and thus, they are less likely to benefit from it.

The Role of Prior Domain Knowledge

Experiments with the Fractions Tutor

An experiment with the Fractions Tutor investigated whether the effectiveness of conceptual and perceptual connection-making support depends on students' level of prior domain knowledge. A total of 105 4th- and 5th-grade students worked with the Fractions Tutor during their regular math instruction, spread across consecutive days. Students were drawn from three elementary schools in Pennsylvania, USA. In the given academic year, the 4th- and 5th-grade students in this school district's scores on the mathematics Pennsylvania State Standardized Test were ranked as below basic for 19%, basic for 17.5%, proficient for 32.3%, and advanced for 31.3%.

All students worked through the entire Fractions Tutor curriculum. They were randomly assigned to one of four conditions that differed with respect to whether they included conceptual and perceptual connection-making problems. All conditions received individual-representations problems in which they solved fractions problems with only one graphical representation at a time, as is typical for common textbook problems. The control condition (no-conceptual / no-perceptual support) worked only on individual-representation problems without connection-making support. The conceptual / no-perceptual condition received individual-representations problems and conceptual connection-making problems. The no-conceptual / perceptual condition received individual-representations problems and perceptual connection-making problems. The conceptual / perceptual condition received individual-representations problems, conceptual connection-making problems, and perceptual connection-making problems. The number of problems per condition was chosen so that the number of total steps was equal for all conditions, to ensure that the amount of practice was equal. Individual, conceptual, and perceptual problems were interleaved for each topic covered by the Fractions Tutor curriculum.

Prior to the study, students took a pretest that assessed their conceptual and procedural knowledge about fractions. After they finished their work with the Fractions Tutor, they took isomorphic immediate posttests. One week later, they took isomorphic delayed posttests.

For learning of fractions concepts, the results showed a significant advantage for conceptual connection-making support on the immediate posttest ($p < 0.05$) and the delayed posttests ($p < 0.05$), but no significant advantage for perceptual connection-making support, nor any significant interactions with students' prior knowledge about fractions concepts ($p > 0.10$). For learning of fractions procedures, there were no significant effects on the immediate posttest. There was a significant advantage for conceptual connection-making support on the delayed posttest ($p < 0.05$). This main effect was qualified by a significant interaction of conceptual connection-making support with prior knowledge about fractions procedures ($p < 0.05$), such that students with high prior knowledge about fractions procedures benefited more from receiving conceptual connection-making support than students with low prior knowledge.

In contrast to this experiment, the evaluation of the Fractions Tutor mentioned above (see section 0; Rau et al. 2012) did not find interactions with students' prior knowledge about fractions concepts or fractions procedures. This disparity between the two experiments might be explained by the difference in populations. The comparison of the school district' performance on the standardized tests in the given academic year shows that the interaction between conceptual support and prior knowledge was only present in the population with lower standardized test scores. This observation is in line with the interpretation that some preliminary level of prior domain knowledge is necessary so that students can benefit from conceptual connection-making support.

Experiments with Chem Tutor

An experiment with Chem Tutor shows a different pattern. This experiment investigated whether the effectiveness of conceptual and perceptual connection-making support depends on students' level of prior domain knowledge. A total of 66 undergraduate students were drawn from an introductory chemistry course for science majors. The study took place in the middle of the semester. All students worked with Chem Tutor's bonding unit. They were randomly assigned to one of four conditions that differed with respect to whether they included conceptual and perceptual connection-making problems. All conditions received individual-representations problems in which they solved chemistry problems with only one graphical representation at a time, as is typical for common textbook problems. The control condition (no-conceptual / no-perceptual support) worked on individual-representation problems without connection-making support. The conceptual / no-perceptual condition received individual-representations problems and conceptual connection-making problems. The no-conceptual / perceptual condition received individual-representations problems and perceptual connection-making problems. The conceptual / perceptual condition received individual-representations problems, conceptual connection-making problems, and perceptual connection-making problems. The number of problems per condition was chosen so that the number of total steps was equal for all conditions, to ensure that the amount of practice was equal. Regular, conceptual, and perceptual problems were interleaved for each pair of representations.

Students took three chemistry knowledge tests: a pretest, intermediate posttest, and final posttest. The tests focused on concepts of bonding. They included reproduction and transfer items and items with and without graphical representations. In addition, students' completed a test that assessed their prior conceptual connection-making competence and a test prior that assessed their prior perceptual connection-making competence.

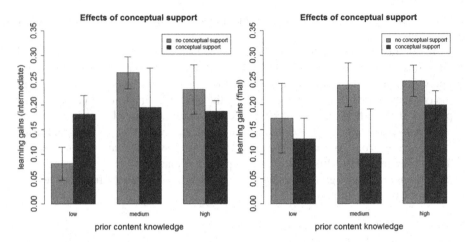

Fig. 8 Effect of conceptual support by prior chemistry knowledge (0-33th, 34th–66th, and 67th–100th percentiles) on learning gains on the intermediate posttest (left) and at the final posttest (right). Error bars show standard errors of the mean

This lab experiment involved two 90 min sessions. In session 1, students first received the pretests and then completed half of the tutor problems. At the end of session 1, they took the intermediate posttest. In session 2, they worked through the remaining tutor problems and then took the final posttest.

The results showed no significant main effects of conceptual support ($p > 0.10$) or perceptual support ($p > 0.10$). Conceptual support interacted significantly with students' prior chemistry knowledge at the intermediate posttest ($p < 0.05$), and marginally significantly at the final posttest, ($p = 0.06$). The left part of Fig. 7 illustrates the nature of this interaction effect for the intermediate posttest, which as was given after students had finished the first half of Chem Tutor's bonding unit. Students with low prior chemistry knowledge learned better with conceptual support at the beginning of the intervention. Students with high prior chemistry knowledge learned better without conceptual support. The right part of Fig. 8 shows that, later in the intervention, all students learned better without conceptual support.

This finding stands in contrast to the experiment mentioned above that evaluated the effectiveness of Chem Tutor (see section 0; Rau and Wu 2015): in that experiment, there were no significant interactions of conceptual support with students' prior chemistry knowledge. Again, this disparity between the two experiments might be explained by the difference in populations. The majority of the students in the evaluation experiment were non-science majors taking an introductory chemistry course. According to the instructor of the non-science majors course, these students had not encountered the content covered in the given Chem Tutor unit before, and they were unfamiliar with most of the graphical representations used. By contrast, students in this second experiment were enrolled in a course for science majors, and—according to their instructor—they were familiar with basic bonding concepts and had seen all of the graphical representations used in the given tutor unit before. Thus, we can assume that students in the second experiment generally had higher prior domain knowledge than students in the evaluation experiment. These observations suggest that students at more advanced levels, conceptual connection-making support is no longer effective.

Discussion

It is difficult to make comparisons across the experiments with the Fractions Tutor and Chem Tutor. The populations were different: the Fractions Tutor was tested with elementary-school students; Chem Tutor was tested with undergraduate students. These populations differ with respect to their developmental level, learning motivation, and also with respect to their diversity—students who have learned together in the classroom, potentially with the same math teacher over several years, are arguably more homogenous than first-semester undergraduate students who came to the university from different states. The domains were different: the Fractions Tutor covers early math topics; Chem Tutor covers advanced science topics. The settings were different: the Fractions Tutor experiments were carried out in school classrooms; the experiments with Chem Tutor were carried out in the lab or online as part of a homework assignment. The duration of the interventions were different too: the Fractions Tutor experiment took 10 h over the course of several consecutive school days; the Chem Tutor experiments took 3 h over the course of no more than 3 days.

Yet, we may gain some interesting insights by comparing the findings obtained across these vastly different populations, topics, and settings. The pattern that emerges across the experiments with the Fractions Tutor and Chem Tutor suggest that (1) conceptual connection-making support is more effective for students who have some basic domain knowledge (Rau and Wu 2015; Rau et al. 2012), and (2) conceptual connection-making support is less effective for students with proficient levels of domain knowledge.

The Role of Sequencing Conceptual and Perceptual Connection-Making Support

When we combine conceptual and perceptual connection-making support, we need to decide on a sequence in which we provide these types of support. This bears the question: Does one competence build on the other? If that is the case, then we should sequence support for these competences accordingly, and ensure that the student has acquired the prerequisite connection-making competence before providing support for the other.

On the one hand, one might argue that perceptual connection-making competences enhance students' benefit from sense-making problems. If this is true, we expect that students will learn best if they become perceptually fluent in making connections before they learn to make sense of connections conceptually. Students who are fluent in making connections based on perceptual features may benefit from increased cognitive capacity during subsequent learning tasks (Kellman et al. 2009; Koedinger et al. 2012). Thus, they can invest more cognitive resources in understanding the conceptual nature of connections and to reason about domain-relevant concepts. Based on these conjectures, providing perceptual connection-making problems before conceptual connection-making problems might decrease the risk of cognitive overload while students work on the conceptual problems, which is known to hamper learning (Chandler and Sweller 1991). This hypothesis is in line with studies that indicate that students perform better on domain knowledge tests after having worked on perceptual connection-making problems (Kellman et al. 2008, 2009).

On the other hand, one might argue that conceptual connection-making competences enhance students' benefit from perceptual connection-making problems. If this is true, students should show the best learning outcomes if they first learn to conceptually understand connections before they become perceptually fluent in making them. Research on conceptual connection-making competences shows that students have difficulties in making connections at a conceptual level and typically do not make them spontaneously (Ainsworth et al. 2002; Rau et al. 2014a, b). Therefore, students may not be able to discover what features of the representations depict meaningful information while working on perceptual connection-making problems. Not having conceptual understanding might lead students to employ inefficient learning strategies (e.g., trial and error), which might impede their benefit from perceptual connection-making problems. Indeed, participants in Kellman and colleagues' studies were typically not novices, but had some considerable prior knowledge about the domain-relevant concepts (e.g., Kellman et al. 2008, 2009). Thus, conceptual understanding of connections might equip students with the knowledge they need in order to attend to relevant features of the graphical representations while they work on perceptual connection-making problems. This hypothesis is in line with an implicit assumption that many educational practice guides seem to make. These practice guides typically provide "checklists" of knowledge that students should have acquired by specific grade levels. Generally, conceptual understanding is expected before perceptual fluency. For example, the NCTM (2006) expects conceptual understanding of fractions representations by the end of grade 5. The ability to efficiently work with fractions representations is expected at the end of grade 8.

Experiment with the Fractions Tutor

An experiment with the Fractions Tutor compared different sequences of conceptual and perceptual connection-making support. A total of 74 elementary-school students participated in the experiment (Rau et al. 2014a, b). Students were recruited through advertisements in local newspapers, online bulletin boards, and through flyers distributed in local schools. Experiment sessions took 1.5 h and were conducted in the lab. Students were randomly assigned to work with one of two different versions of the Fractions Tutor. Students in the conceptual-perceptual condition worked with a version of the tutor that provided conceptual connection-making problems before perceptual connection-making problems. Students in the perceptual-conceptual condition received perceptual connection-making problems before conceptual connection-making problems. Students' learning outcomes were assessed based on a perceptual connection-making test, a conceptual connection-making test, and a fractions knowledge test. In addition, interviews with students were used to assess the quality of their conceptual reasoning about fractions. Furthermore, students' problem-solving performance was assessed based on errors they make while solving the tutor problems.

With respect to the learning outcome measures, there was a significant advantage of the conceptual-perceptual condition over the perceptual-conceptual condition on the perceptual connection-making test ($p < 0.05$). There were no significant differences on the conceptual connection-making test ($p > 0.10$). There was a marginally significant advantage of the conceptual-perceptual condition over the perceptual-conceptual condition on the fractions knowledge test ($p < 0.10$). An analysis of the student interview data showed that students in the conceptual-perceptual condition made more utterances that were coded as high-quality conceptual reasoning about fractions than students in the perceptual-conceptual condition ($p < 0.05$).

The analysis of students' problem-solving performance based on the tutor logs showed that students in the conceptual-perceptual condition made marginally significantly fewer errors on perceptual connection-making problems than students in the perceptual-conceptual condition ($p < 0.10$). That is, students who received conceptual problems before perceptual problems make fewer errors on perceptual problems compared to students who did not receive conceptual problems before. By contrast, students in the perceptual-conceptual condition made marginally significantly more problems on conceptual connection-making problems than students in the conceptual-perceptual condition ($p < 0.10$). That is, students who had received perceptual problems before conceptual problems made more errors on conceptual problems compared to students who had not received perceptual problems before. A follow-up mediation analysis showed that these differences in students' problem-solving performance mediate the advantage of the conceptual-perceptual condition on the fractions knowledge test. In other words, the fact that students in the conceptual-perceptual condition showed lower error rates on perceptual problems explains their better performance on the fractions knowledge test.

Taken together, these findings indicate that conceptual connection-making competences enhance students' benefit from perceptual connection-making support more so than vice versa. The fact that most of the differences between conditions were only marginally significant warrants further investigation. However, the fact that the same differences were found on a number of dependent measures lends credibility to the overall interpretation that perceptual connection-making competences build on conceptual connection-making competences.

Experiment with Chem Tutor

An experiment with Chem Tutor investigated the effects of sequencing conceptual and perceptual support in a factorial design. They were recruited from an introductory chemistry course for science majors. The study took place towards the end of the semester. Students accessed all materials online. Students were randomly assigned to one of five conditions: (1) no-conceptual / no-perceptual support, (2) conceptual / no-perceptual support, (3) no-conceptual / perceptual support, (4) conceptual-then-perceptual support, (5) perceptual-then-conceptual support. Students took three chemistry knowledge tests: a pretest, intermediate posttest, and final posttest. The tests focused on concepts of bonding. They included reproduction and transfer items and items with and without graphical representations. In addition, students' completed a test that assessed their prior conceptual connection-making competence and a test prior that assessed their prior perceptual connection-making competence.

Results showed no significant main effects of conceptual support ($p > 0.10$) or perceptual support ($p > 0.10$). The effect of sequence of conceptual and perceptual support was significant at the final posttest ($p < 0.05$), but it was qualified by a significant interaction with prior conceptual-connections knowledge at the intermediate posttest ($p < 0.05$), and at the final posttest ($p < 0.01$). Other interactions were not significant ($p > 0.10$). Fig. 9 illustrates the interaction of sequence with prior conceptual-connections knowledge. These findings show that the student's prior

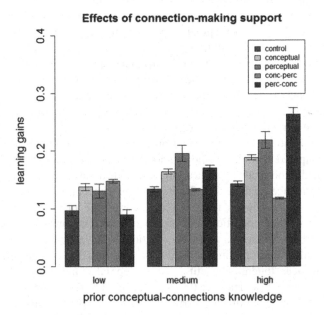

Fig. 9 Effect of connection-making support on gains at final posttest by prior conceptual-connections knowledge (0-33th, 34th–66th, and 67th–100th percentiles). Error bars show standard errors of the mean

conceptual-connections knowledge determines which combination of conceptual and perceptual support is most effective. Students with low prior conceptual-connections knowledge benefited from the combination *only if* conceptual support was provided before perceptual support. If they received perceptual before conceptual support, they did not outperform the control condition. By contrast, students with high prior conceptual-connections knowledge benefited from the combination *only if* perceptual support was provided before conceptual support. If they received conceptual before perceptual support, they performed no better than the control condition. Finally, students with medium prior conceptual-connections knowledge benefited most from perceptual support alone.

Discussion

Again, even though comparisons across the experiments with the Fractions Tutor and Chem Tutor are highly speculative, they may yield some interesting observations that are worth exploring in future research. In contrast to the experiment with the Fractions Tutor, the experiment with Chem Tutor suggests that the interplay between conceptual and perceptual connection-making competences is not a one-way street, but that one benefits the other. However, an interesting parallel between these two experiments is that the findings from the Fractions Tutor—that perceptual connection-making competence builds on conceptual connection-making competence—is in line with the conclusion from the experiment with Chem Tutor: the effectiveness of perceptual support depends on students having previously received conceptual support or students having medium or high prior conceptual-connections competence.

An important contrast between the two experiments is that the experiment with Chem Tutor shows that for students with high prior conceptual-connections competence, providing perceptual before conceptual support is most effective. As mentioned above, many factors may have contributed to different findings in these different populations, topics, and settings, so that it is impossible to draw definite conclusions post hoc. However, the results demonstrate that combining conceptual and perceptual connection-making support is not straightforward: the effectiveness of combining the two different types of support depends on how we sequence them, and different sequences appear to be effective for different types of students.

Preliminary Principles for Adaptive Support and Future Research

The experiments summarized above indicate that combining conceptual and perceptual connection-making support is effective in enhancing students' learning of the domain knowledge. However, the effectiveness of different combinations depends on a number of student characteristics. The following preliminary principles emerge:

Support for perceptual connection-making competences is most effective for students with high mental rotation ability This principle is based on only one experiment that tested mental rotation ability, so more research is needed to investigate the role of mental rotation ability in students' acquisition of perceptual connection-making competences in other domains and with other populations. It seems plausible that educational technologies might be more effective if they adapted to an individual student's mental rotation ability by, for example, providing mental rotation trainings before students work on perceptual connection-making problems, or by providing easier perceptual problems in which the representations are spatially more aligned. However, these hypotheses remain to be tested empirically.

Support for conceptual connection-making competences is most effective for students who have a basic level of prior domain knowledge but who are not yet fully proficient This finding suggests that connection-making support might be most effective if it adapts the time at which conceptual connection-making support is provided to an individual student to the student's level of domain knowledge. More research is needed to investigate what level of domain knowledge is necessary for conceptual connection-making support to be maximally effective. Furthermore, adaptive technologies should fade out the amount of conceptual connection-making support when students reach a high level of proficiency at the domain knowledge. Again, more research is needed to investigate when conceptual connection-making support starts to lose its effectiveness.

The sequence in which conceptual and perceptual connection-making support should be provided depends on the students' level of conceptual connection-making competences It seems that a combination of conceptual-then-perceptual support is effective for students with low levels of conceptual connection-making competence. Thus, adaptive educational technologies may be effective if they start to fade-in perceptual connection-making support towards the beginning of the learning experience. As students gain a medium level of conceptual connection-making competence, they seem to benefit most from receiving only perceptual connection-making support. Thus, adaptive educational technologies may fade-out conceptual connection-making support when a medium level of conceptual connection-making competence is achieved. Students with high prior conceptual connection-making competence seem to benefit from a sequence of perceptual-then-conceptual support. Thus, adaptive educational technologies may fade-in conceptual support later during the learning experience, now providing it after conceptual connection-making support. Future research should investigate how best to identify the point at which conceptual and perceptual support should be faded in or out. Long-term studies would be suitable to address this question, because the experiments summarized in this chapter were exclusively cross-sectional experiments: they investigated the effects of students' prior competences as they were at the beginning of the intervention, and they did not test whether the same students would benefit from a different sequence of conceptual and perceptual support later during the learning experience.

Conclusions

I discussed experiments that extend prior research on conceptual connection-making competences and perceptual connection-making competences, and that present a step towards closing the gap between (so far) separate lines of research that has focused on only one of these competences at a time. The results described in this chapter illustrate that we have much to gain from closing this gap. Because both types of connection-making competences play an important role in STEM learning (Even 1998; Kellman and Massey 2013; Kozma and Russell 2005; Pape and Tchoshanov 2001; Wu et al. 2001), instructors and designers of educational technologies might be inclined to combine support for them. Yet, without principle-based guidance, they might do so in the wrong way: results from both experiments showed that "naïve" combinations of conceptual and perceptual support may not help students but lead to lower learning outcomes than the state of the art. Only if we combine conceptual and perceptual support carefully, by taking into account different aspects of their prior competences in a principled fashion, we may help students make connections and thereby significantly enhance learning outcomes in STEM domains.

Unfortunately, the results indicate that combining conceptual and perceptual connection-making support is not straightforward. There is a need to investigate the nature of complex interactions between prior knowledge about content and connections with students' benefit from conceptual and perceptual connection-making support. Conducting such studies across domains and across student populations that differ in context and developmental level may be a promising route to tease apart which factors contribute to this complex interaction.

Yet, a (preliminary) overarching story across these findings may be that students' acquisition of perceptual connection-making competences builds conceptual connection-making competences, at least at the beginning of an intervention. This finding appears to be somewhat robust across different domains (chemistry vs. math), populations (undergraduate vs. elementary-school students), and educational settings (lab vs. classroom). Supporting students in making connections is of broad relevance because multiple graphical representations are pervasive in STEM domains. Consequently, these findings have the potential to impact a broad range of educational technologies for STEM learning.

References

Ainsworth, S. (2006). Deft: A conceptual framework for considering learning with multiple representations. *Learning and Instruction, 16*, 183–198.

Ainsworth, S. (2008a). How should we evaluate multimedia learning environments? *Understanding multimedia documents* (pp. 249–265).

Ainsworth, S. (2008b). How do animations influence learning? In D. H. Robinson & G. Schraw (Eds.), *Current perspectives on cognition, learning, and instruction: Recent innovations in educational technology that facilitate student learning* (pp. 37–67). Charlotte: Information Age Publishing.

Ainsworth, S. (2008c). The educational value of multiple-representations when learning complex scientific concepts. In J. K. Gilbert, M. Reiner, & A. Nakama (Eds.), *Visualization: Theory and Practice in Science Education* (pp. 191–208). Netherlands: Springer.

Ainsworth, S., Bibby, P., & Wood, D. (2002). Examining the effects of different multiple representational systems in learning primary mathematics. *Journal of the Learning Sciences, 11*, 25–61.

Arcavi, A. (2003). The role of visual representations in the learning of mathematics. *Educational Studies in Mathematics, 52*, 215–241.

Berthold, K., & Renkl, A. (2009). Instructional aids to support a conceptual understanding of multiple representations. *Journal of Educational Research, 101*(1), 70–87.

Berthold, K., Eysink, T. H. S., & Renkl, A. (2008). Assisting self-explanation prompts are more effective than open prompts when learning with multiple representations. *Instructional Science, 27*, 345–363.

Betrancourt, M. (2005). The animation and interactivity principles in multimedia Learning. In R. E. Mayer (Ed.), *The Cambridge handbook of multimedia learning* (pp. 287–296). New York: Cambridge University Press.

Bodemer, D., & Faust, U. (2006). External and mental referencing of multiple representations. *Computers in Human Behavior, 22*, 27–42.

Bodemer, D., Ploetzner, R., Feuerlein, I., & Spada, H. (2004). The active integration of information during learning with dynamic and interactive visualisations. *Learning and Instruction, 14*, 325–341.

Bodemer, D., Ploetzner, R., Bruchmüller, K., & Häcker, S. (2005). Supporting learning with interactive multimedia through active integration of representations. *Instructional Science, 33*, 73–95.

Chandler, P., & Sweller, J. (1991). Cognitive load theory and the format of instruction. *Cognition and Instruction, 8*, 293–332.

Charalambous, C. Y., & Pitta-Pantazi, D. (2007). Drawing on a theoretical model to study students' understandings of fractions. *Educational Studies in Mathematics, 64*, 293–316.

Chi, M. T. H., Feltovich, P. J., & Glaser, R. (1981). Categorization and representation of physics problems by experts and novices. *Cognitive Science, 5*, 121–152.

Cook, M., Wiebe, E. N., & Carter, G. (2007). The influence of prior knowledge on viewing and interpreting graphics with macroscopic and molecular representations. *Science Education, 92*, 848–867.

Corbett, A. T., Koedinger, K., & Hadley, W. S. (2001). Cognitive tutors: From the research classroom to all classrooms. In P. S. Goodman (Ed.), *Technology enhanced learning:Opportunities for change* (pp. 235–263). Mahwah: Lawrence Erlbaum Associates Publishers.

Cramer, K. (2001). Using models to build an understanding of functions. *Mathematics Teaching in the Middle School, 6*, 310–318.

Dori, Y. J., & Barak, M. (2001). Virtual and physical molecular modeling: Fostering model perception and spatial understanding. *Educational Technology & Society, 4*, 61–74.

Dreyfus, H., & Dreyfus, S. E. (1986). *Five steps from novice to expert mind over machine: The power of human intuition and expertise in the era of the computer* (pp. 16–51). New York: The Free Press.

Eilam, B., & Poyas, Y. (2008). Learning with multiple representations: Extending multimedia learning beyond the lab. *Learning and Instruction, 18*, 368–378.

Even, R. (1998). Factors involved in linking representations of functions. *The Journal of Mathamtical Behavior, 17*, 105–121.

Gegenfurtner, A., Lehtinen, E., & Säljö, R. (2011). Expertise differences in the comprehension of visualizations: A meta-analysis of eye-tracking research in professional domains. *Educational Psychology Review, 23*, 523–552.

Gentner, D. (1983). Structure-mapping: A theoretical framework for analogy. *Cognitive Science, 7*, 155–170.

Gibson, E. J. (1969). *Principles of perceptual learning and development*. New York: Prentice Hall.

Gibson, E. J. (2000). Perceptual learning in development: Some basic concepts. *Ecological Psychology, 12*, 295–302.

Gilbert, J. K. (2008). Visualization: An emergent field of practice and inquiry in science education. In J. K. Gilbert, M. Reiner, & M. B. Nakhleh (Eds.), *Visualization: Theory and practice in science education* (pp. 3–24). Dordrecht: Springer.

Gutwill, J. P., Frederiksen, J. R., & White, B. Y. (1999). Making their own connections:Students' understanding of multiple models in basic electricity. *Cognition and Instruction, 17*, 249–282.

Hegarty, M., & Waller, D. A. (2005). Individual differences in spatial abilities. In P. Shah & A. Miyake (Eds.), *The Cambridge handbook of visuospatial thinking* (pp. 121–169). New York: Cambridge University Press.

Holzinger, A., Kickmeier-Rust, M. D., & Albert, D. (2008). Dynamic media in computer science education; Content complexity and learning performance: Is less more? *Educational Technology & Society, 11*, 279–290.

Jones, L. L., Jordan, K. D., & Stillings, N. A. (2005). Molecular visualization in chemistry education: The role of multidisciplinary collaboration. *Chemistry Education Research and Practice, 6*, 136–149.

de Jong, T., & van Joolingen, W. R. (1998). Scientific discovery learning with computer simulations of conceptual domains. *Review of Educational Research, 68*, 179–201.

de Jong, T., Ainsworth, S. E., Dobson, M., Van der Meij, J., Levonen, J., & Reimann, P. (1998). Acquiring knowledge in science and mathematics: The use of multiple representations in technology-based learning environments. In M. W. Van Someren, W. Reimers, H. P. A. Boshuizen, & T. de Jong (Eds.), *Learning with Multiple Representations* (pp. 9–41). Bingley: Emerald Group Publishing Limited.

Kellman, P. J., & Garrigan, P. B. (2009). Perceptual learning and human expertise. *Physics of Life Reviews, 6*, 53–84.

Kellman, P. J., & Massey, C. M. (2013). Perceptual learning, cognition, and expertise. *The psychology of learning and motivation, 558*, 117–165.

Kellman, P. J., Massey, C. M., Roth, Z., Burke, T., Zucker, J., Saw, A., .Wise, J. (2008).Perceptual learning and the technology of expertise: Studies in fraction learning and algebra. *Pragmatics & Cognition, 16*, 356-405.

Kellman, P. J., Massey, C. M., & Son, J. Y. (2009). Perceptual learning modules in mathematics: Enhancing students' pattern recognition, structure extraction, and fluency. *Topics in Cognitive Science, 1*, 285–305.

Koedinger, K. R., & Aleven, V. (2007). Exploring the assistance dilemma in experiments with cognitive tutors. *Educational Psychology Review, 19*, 239–264.

Koedinger, K. R., & Corbett, A. (2006). *Cognitive tutors: Technology bringing learning sciences to the classroom*. New York: Cambridge University Press.

Koedinger, K. R., Corbett, A. T., & Perfetti, C. (2012). The knowledge-learning-instruction framework: Bridging the science-practice chasm to enhance robust student learning. *Cognitive Science, 36*, 757–798.

Kordaki, M. (2010). A drawing and multi-representational computer environment for beginners' learning of programming using C: Design and pilot formative evaluation. *Computers & Education, 54*, 69–87.

Kozma, R., & Russell, J. (2005). Students becoming chemists: Developing representational competence. In J. Gilbert (Ed.), *Visualization in science education* (pp. 121–145). Dordrecht: Springer.

Kozma, R., Chin, E., Russell, J., & Marx, N. (2000). The roles of representations and tools in the chemistry laboratory and their implications for chemistry learning. *The Journal of the Learning Sciences, 9*, 105–143.

Larkin, J. H., & Simon, H. A. (1987). Why a diagram is (sometimes) worth ten thousand words. *Cognitive Science: A Multidisciplinary Journal, 11*, 65–100.

Lewalter, D. (2003). Cognitive strategies for learning from static and dynamic visuals. *Learning and Instruction, 13*, 177–189.

Linenberger, K. J., & Bretz, S. L. (2012). Generating cognitive dissonance in student interviews through multiple representations. *Chemistry Education Research and Practice, 13*, 172–178.

Massey, C. M., Kellman, P. J., Roth, Z., & Burke, T. (2011). Perceptual learning and adaptive learning technology - developing new approaches to mathematics learning in the classroom. In N. L. Stein & S. W. Raudenbush (Eds.), *Developmental cognitive science goes to school* (pp. 235–249). New York: Routledge.

Mayer, R. E. (2003). The promise of multimedia learning: Using the same instructional design methods across different media. *Learning and Instruction, 13*, 125–139.

van der Meij, J., & de Jong, T. (2006). Supporting students' learning with multiple representations in a dynamic simulation-based learning environment. *Learning and Instruction, 16*, 199–212.

Van der Meij, J., & de Jong, T. (2011). The effects of directive self-explanation prompts to support active processing of multiple representations in a simulation-based learning environment. *Journal of Computer Assisted Learning, 27*, 411–423.

Moss, J. (2005). Pipes, tubes, and beakers: New approaches to teaching the rational-number system. In J. Brantsford & S. Donovan (Eds.), *How people learn: A targeted report for teachers* (pp. 309–349). Washington, D.C.: National Academy Press.

NCTM. (2000). *Principles and standards for school mathematics*. Reston: National Council of Teachers of Mathematics..

NCTM. (2006). *Curriculum focal points for prekindergarten through grade 8 mathematics: A quest for coherence*. VA: Reston.

Özgün-Koca, S. A. (2008). Ninth grade students studying the movement of fish to learn about linear relationships: The use of video-based analysis software in mathematics classrooms. *The Mathematics Educator, 18*, 15–25.

Pape, S. J., & Tchoshanov, M. A. (2001). The role of representation (s) in developing mathematical understanding. *Theory into Practice, 40*, 118–127.

Patel, Y., & Dexter, S. (2014). Using multiple representations to build conceptual understanding in science and mathematics. In M. Searson & M. Ochoa (Eds.), *Proceedings of society for information technology & teacher education international conference 2014* (pp. 1304–1309). Chesapeake: AACE.

Peters, M., Laeng, B., Latham, K., Jackson, M., Zaiyouna, R., & Richardson, C. (1995). A redrawn Vandenberg & Kuse mental rotations test: Different versions and factors that affect performance. *Brain and Cognition, 28*, 39–58.

Rau, M. A., & Evenstone, A. L. (2014). Multi-methods approach for domain-specific grounding: An ITS for connection making in chemistry. In S. Trausan-Matu, K. E. Boyer, M. Crosby & K. Panourgia (Eds.), Proceedings of the 12th International conference on intelligent tutoring systems (pp. 426–435). Berlin/Heidelberg: Springer.

Rau, M. A., & Wu, S. P. W. (2015). ITS support for conceptual and perceptual processes in learning with multiple graphical representations. In C. Conati, N. Heffernan, A. Mitrovic, & M. F. Verdejo (Eds.), *Artificial intelligence in education* (pp. 398–407). Switzerland: Springer International Publishing.

Rau, M. A., Aleven, V., Rummel, N., & Rohrbach, S. (2012). Sense making alone doesn't do it: Fluency matters too! Its support for robust learning with multiple representations. In S. Cerri,

W. Clancey, G. Papadourakis, & K. Panourgia (Eds.), *Intelligent tutoring systems* (pp. 174–184). Berlin: Springer.

Rau, M. A., Aleven, V., Rummel, N., & Rohrbach, S. (2013). Why interactive learning environments can have it all: Resolving design conflicts between conflicting goals. In *Proceedings of the SIGCHI 2013 ACM conference on human factors in computing systems* (pp. 109–118). New York: ACM.

Rau, M. A., Aleven, V., & Rummel, N. (2014a). Sequencing sense-making and fluency-building support for connection making between multiple graphical representations. In J. L. Polman, E. A. Kyza, D. K. O'Neill, I. Tabak, W. R. Penuel, A. S. Jurow, K. O'Connor, T. Lee, & L. D'Amico (Eds.), *Learning and becoming in practice: The international conference of the learning sciences (ICLS 2014)* (pp. 977–981). Boulder: International Society of the Learning Sciences.

Rau, M. A., Aleven, V., Rummel, N., & Pardos, Z. (2014b). How should intelligent tutoring systems sequence multiple graphical representations of fractions? A multi-methods study. *International Journal of Artificial Intelligence in Education, 24*, 125–161.

Rau, M. A., Michaelis, J. E., & Fay, N. (2015). Connection making between multiple graphical representations: A multi-methods approach for domain-specific grounding of an intelligent tutoring system for chemistry. *Computers and Education, 82*, 460–485.

Richman, H. B., Gobet, F., Staszewski, J. J., & Simon, H. A. (1996). Perceptual and memory processes in the acquisition of expert performance: The epam model. In K. A. Ericsson (Ed.), *The road to excellence? The acquisition of expert performance in the arts and sciences, sports and games* (pp. 167–187). Mahwah: Erlbaum Associatees.

Schnotz, W. (2005). An integrated model of text and picture comprehension. In R. E. Mayer (Ed.), *The Cambridge handbook of multimedia learning* (pp. 49–69). New York: Cambridge University Press.

Schnotz, W., & Bannert, M. (2003). Construction and interference in learning from multiple representation. *Learning and Instruction, 13*, 141–156.

Schwonke, R., Renkl, A., Salden, R., & Aleven, V. (2011). Effects of different ratios of worked solution steps and problem solving opportunities on cognitive load and learning outcomes. *Computers in Human Behavior, 27*, 58–62.

Seufert, T. (2003). Supporting coherence formation in learning from multiple representations. *Learning and Instruction, 13*, 227–237.

Seufert, T., & Brünken, R. (2006). Cognitive load and the format of instructional aids for coherence formation. *Applied Cognitive Psychology, 20*, 321–331.

Siegler, R. S., Carpenter, T., Fennell, F., Geary, D., Lewis, J., Okamoto, Y., . . . Wray, J. (2010). *Developing effective fractions instruction: A practice guide*. Washington, DC: National Center for Education Evaluation and Regional Assistance, Institute of Education Sciences, U.S. Department of Education.

Stern, E., Aprea, C., & Ebner, H. G. (2003). Improving cross-content transfer in text processing by means of active graphical representation. *Learning and Instruction, 13*, 191–203.

Stieff, M. (2007). Mental rotation and diagrammatic reasoning in science. *Learning and Instruction, 17*, 219–234.

Taber, S. B. (2001). Making connections among different representations: The case of multiplication of fractions. Paper presented at the Annual meeting of the American Educational Research Association (Seattle, WA, April 10–14, 2001).

Talanquer, V. (2013). Chemistry education: Ten facets to shape us. *Journal for Research in Mathematics Education, 90*, 832–838.

Uttal, D. H., Meadow, N. G., Tipton, E., Hand, L. L., Alden, A. R., Warren, C., & Newcombe, N. S. (2013). The malleability of spatial skills: A meta-analysis of training studies. *Psychological Bulletin, 139*, 352–402.

Van Labeke, N., & Ainsworth, S. E. (2002). Representational decisions when learning population dynamics with an instructional simulation. In S. A. Cerri, G. Gouardères & F. Paraguacu

(Eds.), *Proceedings of the 6th international conference intelligent tutoring systems* (pp. 831–840): Springer Verlag.

Van Someren, M. W., Boshuizen, H. P. A., & de Jong, T. (1998). Multiple representations in human reasoning. In M. W. Van Someren, H. P. A. Boshuizen, & T. de Jong (Eds.), *Learning with multiple representations* (pp. 1–9). Pergamon: Oxford.

VanLehn, K. (2011). The relative effectiveness of human tutoring, intelligent tutoring systems and other tutoring systems. *Educational Psychologist, 46,* 197–221.

Vreman-de Olde, C., & De Jong, T. (2007). Scaffolding learners in designing investigation assignments for a computer simulation. *Journal of Computer Assisted Learning, 22,* 63–73.

Wai, J., Lubinski, D., & Benbow, C. P. (2009). Spatial ability for stem domains: Aligning over 50 years of cumulative psychological knowledge solidifies its importance. *Journal of Educational Psychology, 101,* 817–835.

Wise, J. A., Kubose, T., Chang, N., Russell, A., & Kellman, P. J. (2000). Perceptual learning modules in mathematics and science instruction. In P. Hoffman & D. Lemke (Eds.), *Teaching and learning in a network world* (pp. 169–176). Amsterdam: IOS Press.

Wu, H. K., & Shah, P. (2004). Exploring visuospatial thinking in chemistry learning. *Science Education, 88*(3), 465–492.

Wu, H. K., Krajcik, J. S., & Soloway, E. (2001). Promoting understanding of chemical representations: Students' use of a visualization tool in the classroom. *Journal of Research in Science Teaching, 38,* 821–842.

Zhang, J. (1997). The nature of external representations in problem solving. *Cognitive Science, 21,* 179–217.

Zhang, J., & Norman, D. A. (1994). Representations in distributed cognitive tasks. *Cognitive Science: A Multidisciplinary Journal, 18,* 87–122.

Instructional Representations as Tools to Teach Systems Thinking

Tammy Lee and Gail Jones

Representational Competence and Systems Thinking

Creating scientifically literate citizens who engage in public debate and participate in decision-making processes concerning complex scientific issues is a major goal of science education. The complexity of systems and the development of systems thinking in science is one of the multifaceted issues of science that is increasingly impacting us in our advancing global society. As our knowledge of science advances, there is an increasing need to understand complex systems, which includes (often dynamic) phenomena and their interrelationships (Hmelo-Silver and Pfeffer 2004).

To be scientifically literate requires an individual to be able to read, write, and communicate the language of science (Krajcik and Sutherland 2010; Norris and Phillips 2003; Yore et al. 2007). But the communication of science involves more than verbal discourse or written text. Science is multimodal which includes communicating with a variety of representations (e.g., graphs, diagrams, symbols, formulae, and pictorial). To effectively communicate this multimodal language, a student needs to interpret, construct, transform, and evaluate different scientific representations in order to conceptually understand science (Kress et al. 2001; Lemke 2004; Yore and Hand 2010). These skills help to build representational competence (Kozma, Chin, Russell, & Marx, Kozma et al. 2000; Kozma and Russell 1997, 2005) and contribute to becoming a scientifically literate individual.

Teaching and learning science in the K-12 classroom requires the use of a variety of external scientific representations (Ainsworth 2006; Kress et al. 2001; Lemke

T. Lee
East Carolina University, Greenville, NC, USA
e-mail: leeta@ecu.edu

G. Jones (✉)
North Carolina State University, Raleigh, NC, USA
e-mail: gail_jones@ncsu.edu

© Springer International Publishing AG, part of Springer Nature 2018   133
K. L. Daniel (ed.), *Towards a Framework for Representational Competence in Science Education*, Models and Modeling in Science Education 11,
https://doi.org/10.1007/978-3-319-89945-9_7

2004; Yore and Hand 2010). Developing students' abilities to use and reason with these scientific representations is essential for learning science and developing representational competence. The study of systems in science necessitates the use of multiple representations due to their complexity in terms of scale, hidden dimensions, and the interplay of relationships among the components and processes of systems. Classroom instruction of systems and the development of systems thinking are dependent upon the selection, interpretation, explanation, and use of effective representations by teachers. This chapter discusses the importance of representational competence in the development of systems thinking among students and provides recommendations for the explicit use of representations when teaching about complex systems.

A Call for Systems Thinking

Systems thinking has been identified as crucial within a number of domains such as social sciences (e.g., Senge 1990), medicine (e.g., Faughman and Elson 1998), psychology (e.g., Emery 1992), curriculum development (e.g., Ben-Zvi Assaraf and Orion 2004), decision making (e.g., Graczyk 1993), project management (e.g., Lewis 1998), engineering (e.g., Fordyce 1988) and mathematics (e.g., Ossimitz 2000). Developing an understanding of the components and relationships that comprise complex systems such as ecosystems, moon phases, or energy transfer requires the ability to apply systems thinking (Evagorou et al. 2009). Research in this area has shown individuals need knowledge about the science domain (i.e., physics or biology) as well as higher-order thinking skills to fully conceptualize complex systems (Frank 2000). Furthermore, research has suggested that there is a link between the development of systems thinking and a conceptual understanding of science (Grotzer and Bell-Basca 2003a). Goldstone and Wilensky (2008) maintain that learning about systems allows for cross-disciplinary inquiry across science areas. Although there is an increasing recognition that all areas of science require systems thinking to critically reason scientifically, there is limited research on how teachers and students develop systems thinking in the context of science education (Kali et al. 2003).

In the past 10 years, science education researchers have begun to recognize the importance of students' abilities to learn about complex systems, and instructional methods and tools being used to teach complex systems. This research includes systems thinking as it relates to technological systems (Frank 2000; Sabelli 2006), social systems (e.g., Booth Sweeney 2000; Booth Sweeney and Sterman 2007; Kim 1999a, b; Mandinach 1989; Steed 1992; Ullmer 1986), biological systems (e.g., Verhoeff et al. 2008), and natural systems (e.g., Ben-Zvi Assaraf and Orion 2005; Hmelo-Silver et al. 2007; Hmelo-Silver and Pfeffer 2004; Ossimitz 2000; Wilensky and Resnick 1999).

Advancement of Technology

Investigating how systems work is not a new process for scientists or engineers; but with advances in technology and the use of modeling, our perspectives of science and the systems within science have been transformed. The advancement of modeling tools allows us to observe systems more closely and develop more precise predictions and inferences about the behaviors of systems. These superior models of natural phenomena are strongly grounded in mathematical and computational reasoning, which expands our knowledge and our ability to understand natural systems, such as climate change (National Research Council 2007). This change in contemporary science has resulted in the increased use of statistical modeling of natural phenomena for visualization of complex systems (Klahr and Simon 1999). Historically, scientists relied on direct causal models; but advancement in scientific and technological understanding of modeling allows scientists to make comparisons of system models to examine system interaction and behavior in numerous scenarios (NRC 2007). Technological advancements have provided us with interdisciplinary information from multiple perspectives, which leads to more accurate predictions regarding system behavior and enhances decision-making. This explosion in technology and availability of these models allows teachers to use these tools in new ways within science instruction.

Call for Reform of the Use of Representations in Teaching Systems Thinking Described in Science Standards

In 1996, *National Science Education Standards* (NSES) introduced the essential components of a system and the *Next Generation of Science Standards* (Achieve 2013) elaborated on the essential aspects of systems for the next generation of science educators:

> The natural and designed world is complex; it is too large and complicated to investigate and comprehend all at once. Scientists and students learn to define small portions for the convenience of investigation. The units of investigations can be referred to as "systems." A system is an organized group of related objects or components that form a whole. Systems can consist, for example, of organisms, machines, fundamental particles, galaxies, ideas, and numbers. Systems have boundaries, components, resources, flow, and feedback. (NRC 2012, pg. 91–92)

Systems and system modeling were included as part of the seven crosscutting concepts identified in the *Framework for K-12 Science Education* (NRC 2012) and NGSS (2013) that provide students with the tools needed to create a more advanced understanding of the disciplinary core ideas of science. The seven crosscutting concepts identified in the Framework and NGSS (2013) include: (1) patterns; (2) cause and effect relationships; (3) scale, proportion and quantity; (4) systems and system models; (5) energy and matter — flows, cycles, and conservation; (6) structure and function; and (7) stability and change. Each of the crosscutting concepts is essential for investigating and understanding natural or designed systems; thus, application of these concepts contributes to the development of a systems thinker. The descriptions of each crosscutting concept explicitly explain how to use the crosscutting concepts to identify components of a system, the interactions of the components of a system, and the systems' behavior as a whole regarding the interrelationships between the components (NRC 2007).

Crosscutting concepts provide students with an organizational framework (NRC 2012) much like a systems thinking "framework" that Senge (1990) described that can assist students with connecting knowledge across various disciplines, creating a coherent and scientifically based view of the world. The framework notes that these crosscutting concepts traditionally have been implicit in nature, hence the rationale for making them their own dimension. Framework developers weaved the crosscutting concepts into performance expectations at all grade levels, ensuring explicit instruction within the various science areas. These concepts also illustrate cross-disciplinary contexts helping students to develop a cumulative, logical, and usable understanding of science and engineering (NRC 2012). If teachers utilize the crosscutting concepts for their intended purpose then the implementation of systems thinking approaches would become a natural component of their instruction.

Visual representations are a major component of early modeling. The practice of developing and using models is one of the eight essential scientific and engineering practices while systems and systems models is one of the seven crosscutting concepts in the NGSS (Achieve 2013). Both scientists and engineers use models, which include sketches, diagrams, mathematical relationships, simulations, and physical models to study systems. These models are used to explore behaviors occurring within the system, make predictions about behavior, as well as make predictions about the relationships among the components of the system. Scientists and engineers use data to assess relationships to determine if revisions should be made in models (NRC 2012). The use of models and representations for the study of systems is essential to the fields of science and engineering.

Developing Systems Thinking and Representational Competence

Although science educators are increasingly emphasizing systems thinking, there are relatively few studies that explore how students learn about complex systems; and teaching complex systems seems to be largely absent from most science classroom curricula (Jacobson and Wilensky 2006). The limited research that exists on the development of systems thinking skills has focused on students of different ages including elementary (Ben-Zvi Assaraf and Orion 2010), middle, high school

(Penner 2000; Frank 2000; Ben-Zvi Assaraf and Orion 2005; Booth Sweeney and Sterman 2007) and college students (Booth Sweeney and Sterman 2007). These studies have suggested that developing a systems thinking approach involves having students examine, evaluate, and invent, which all apply higher-order thinking skills (Frank 2000). Resnick (1987) defined characteristics of higher-order thinking as complex, capable of producing multiple solutions, involving degrees of judgment and uncertainty, utilizing self-regulation, finding structure in apparent disorder, and being fruitful. These skills are comparable to the mental activity involved in systems model building, analysis, and synthesis (Ben-Zvi Assaraf and Orion 2010; Frank 2000).

Some researchers have argued that systems thinking skills include cognitive abilities such as: (a) thinking in terms of dynamic processes, (e.g., delays, feedback loops, oscillations); (b) understanding how the behavior of the system arises from interaction of components over time (e.g., dynamic complexity); (c) discovering and representing feedback processes that underlie observed patterns of the system's behavior; (d) identifying nonlinearities; and (g) scientific thinking, including being able to quantify relations and to hypothesize test assumptions and models (Booth Sweeney 2000; Draper 1993; Frank 2000; Ossimitz 2000). Ossimitz (2000) characterized systems thinking skills as having four central dimensions: (1) network thinking (e.g. thinking in feedback loops); (2) dynamic thinking (e.g. accounting for time lapse); (3) thinking in models (e.g. explicitly comprehend modeling); and (4) system-compatible action (e.g. understand changes in the system). Ossimitz explains that thinking in models not only comprises the selection of models but the ability to build models when interpreting and explaining systems. The last dimension system-compatible action refers to understanding that the system is subject to change, which can impact other parts or components of the system.

One of the key skills for systems thinking is having representational competence. (Nitz et al. 2014) have defined representational competence as key to interpreting, constructing, translating, and evaluating models. Virtually all definitions of systems thinking skills identify the ability to think in terms of models by either knowing how to interpret and build models or understanding the use of models to quantify relations and test assumptions of the system. Representational competence modeling and systems thinking are intertwined skills that have reciprocal influence on each other.

Systems thinking and representational competence are both closely related to students' conceptual understanding of a particular domain of study (Kozma and Russell 1997; Stieff 2011). Kozma and Russell (2005) proposed five levels of representational competence ranging from the novice use of surface-based representations portrayed with symbolic, syntactic, and semantic use of representations to an expert use of representations for reflective and rhetorical purposes. These representational competence developmental levels are not characterized as stage-like or uniform but are reliant on their time and utilization in the classroom. Representational skills can improve when instruction internalizes and integrates representations in physical, symbolic, and social contexts (Kozma and Russell 2005).

Given the need for higher order thinking skills as a component of systems thinking, researchers are not in agreement about the age and cognitive abilities needed to master systems thinking. Evidence indicates that even highly educated adults with extensive mathematics and science backgrounds have poor systems thinking skills (e.g., Booth Sweeney and Sterman 2007; Dorner 1980). Booth Sweeney (2000) used a systems-thinking inventory designed to assess knowledge of concepts, such as feedback, delays, stock and flow relationships and time delay. The study involved students at Massachusetts Institute of Technology business school prior to exposure to system dynamics concepts. The results showed that even highly educated individuals in science had poor levels of understanding in concepts such as stock and flow relationships, time delays, and conservation of matter.

Other researchers, such as (Sheehy et al. 2000), have noted that they had presumed incorrectly that young children could not reach an appropriate level of sophistication of systems thinking until adolescence, the proof of which came after they investigated primary children's understanding of systems thinking in an environmental context, using methodologies that required minimal linguistic input. Through manipulation of simulations, their study revealed that even young students used some systems thinking skills. Although there is a limited amount of research investigating elementary children's development of systems thinking skills (Ben-Zvi Assaraf and Orion 2010), Forrester (2007) argues these skills should be developed early in elementary children since they still have open minds, are inquisitive, and have not been conditioned to see the world in terms of unidirectional cause and effect.

The measure of systems thinking skills and knowledge among students of all ages is still in its infancy. It is not yet clear how teachers develop the pedagogical content knowledge needed to teach systems thinking. Furthermore, how do teachers use representations to teach the complexities of systems that are inherent in science? In the sections that follow the assessments used by researchers to measure systems thinking skills that include representations such as survey instruments, interviews, student-created drawings, concept maps, word associations, and specialized tests measuring specific skills associated with systems thinking are discussed.

Using Representational Competence Abilities to Measure Systems Thinking Skills

Researchers often use representations such as student drawings, concept maps, diagrams, and student-reconstructed diagrams to measure the development of systems thinking skills in elementary and middle school age students (Ben-Zvi Assaraf and Orion 2005, 2010; Kali et al. 2003). In 2010, Ben-Zvi Assaraf and Orion used two types of student drawings to assess elementary students' understanding of the interrelationships between components in the hydrocycle. Students were asked to draw "What happens to water in nature?" Pre- and post-instruction drawings were evaluated to determine the presence of systems thinking based on the following items in their drawings; (a) appearance of processes; (b) appearance of various earth systems; (c) appearance of human consumption or pollution; and (d) cyclic perceptions of the water cycle illustrated by connecting various components. Students' pre-instruction drawings presented the hydrocycle with only atmospheric components (i.e., evaporation, condensation, and precipitation) with the exclusion of groundwater aspects. Post-instruction drawings included the addition of penetration and underground water flow revealing increased knowledge of these groundwater processes. The combination of drawings and interviews showed students had enhanced recognition of the relationships between these components in the subsystems of the hydrocycle.

Ben-Zvi Assaraf and Orion (2010) used the Ecology System Inventory (ESI), to assess elementary students' perceptions of the hidden dimensions of the hydrosphere system (e.g., processes that take place under the surface) after outdoor instruction. The ESI presented students with an image of an ecological system and asked to identify the components, relationships, and hidden dimensions of the hydrosphere. Students were asked to represent these components of the system by drawing them on the ESI. The results of the study showed the ESI was effective in measuring changes in systems knowledge.

Concept maps have long been used as a powerful research tool to examine how learners restructure knowledge (Martin et al. 2000; Mason 1992; Novak and Gowin 1984; Roth 1994). In particular, concept maps have been used to assess the development of systems thinking skills. Ben-Zvi Assaraf and Orion (2005) used concept maps pre- and post-instruction and found middle school students improved skills in: (1) identifying the system's components and processes (i.e. number of concepts); (2) recognizing the appearance of the earth's systems; (3) identifying dynamic relationships within the system (i.e. number of linkages); (4) identifying the appearance of the human aspect; (5) identifying the appearance of cyclic perception of the water cycle; and (6) organizing components and placing them within a framework of relationships. The use of concept maps provides another form of representation that can be used for assessing the development of systems thinking skills.

Kali et al. (2003) used a reconstructed diagram to measure students' knowledge of the dynamic cyclic relationship of the transportation of earth materials within the complex system that includes the rock cycle. Students were engaged in an inquiry lab in the classroom that modeled the effect of transportation of pebbles in a river determined by the shape and size of the pebbles. As a follow-up activity, students used a diagram to hypothesize about the transportation of various pebbles found in the local creek. During a field experience, students verified their hypotheses of rock movement by examining different locations at the river's watershed. Students demonstrated their knowledge of the directional movement of the earth materials through the use of arrows on the reconstructed diagram. Post-test results indicated an increase in understanding the sequences and processes of material transportation as assessed through the use of these reconstructed diagrams.

These examples illustrate the use of representational competence in the measurement of systems thinking skills. For example, students used representations to describe scientific concepts, either by illustration or by adding more words and links to concept maps. Interviews revealed students' abilities to explain their drawings and the representations' appropriateness for a specific purpose when explaining the inclusion of the hidden dimensions from their pre- and post-drawings. In the above instances, students constructed and revised representations to indicate their knowledge of systems thinking. The ability to use a representation to identify, describe and explain a scientific concept and use a representation to support claims and draw inferences were described by Kozma and Russell (2005) as specific skills for demonstrating representational competence. Each of these studies demonstrates how researchers have used representations to measure systems thinking development as well as to promote representational competence.

Theoretical Models Used for Evaluating Systems Thinking

One of the first challenges to teaching and assessing systems thinking is defining and modeling the process. Ben-Zvi Assaraf and Orion (2005) developed the "Systems Thinking Hierarchical Model" (STH Model) to develop systems thinking while learning about complex systems. Several studies have used this model to assess the development of students' levels of systems thinking.

The Systems Thinking Hierarchical model for developing systems thinking proposed by Ben-Zvi Assaraf and Orion (2005), utilized eight characteristics for developing a hierarchical systems thinking structure within the context of the hydrocycle. These eight characteristics were classified into four hierarchical levels: Level 1 includes the ability to identify the system's components and processes; Level 2 includes the ability to identify relationships between separate components and the ability to identify dynamic relationships between the system's components; Level 3 includes the ability to understand the cyclic nature of systems, the ability to organize components and place them within a network of relationships, and the ability to make generalizations; and Level 4 includes an understanding of the hidden components of the system and the system's evolution in time (i.e. prediction and retrospection). These levels can be used as a structure for instruction with each group of skills serving as a basis for the development of the next higher level of systems thinking. Ben-Zvi Assaraf and Orion (2005) suggested that these levels are hierarchical in the development of systems thinking. They reported that elementary, middle, and high school students did not demonstrate growth in higher levels until they first accomplished the beginning levels.

Ben-Zvi Assaraf and Orion's (2005) results raise questions about possible developmental stages that may exist in the development of systems thinking. For instance, Booth Sweeney and Sterman (2007) claimed that when middle school students are probed with questions such as, "What happens next?", they demonstrated an accurate understanding of interactions between objects and, more importantly, grasped how one thing influences another and how these interrelationships hold the system together. Based on their research, Booth Sweeney and Sterman (2007) suggested that the development of systems thinking may not proceed in an ordered and organized sequence.

Another model used for learning about biological systems is known as the Structure-Behavior-Function (SBF) framework (Hmelo et al. 2000; Hmelo-Silver and Pfeffer 2004). This framework was used in studies investigating how students, adults, and experts understand complex biological systems such as a salt marsh and the human body respiratory system. The SBF knowledge representation framework focuses on causal relations between structure and function, which is essential to understanding complex systems in biological contexts. The SBF has been used to describe structures and functions of a system and how they are related through actions (i.e. behaviors). The SBF representation framework also provides a way to analyze how different levels of structures, behaviors, and functions interact with one another within the overall system.

Evagorou et al. (2009) developed seven skills based on a combination of skills identified in the systems thinking literature (Ben-Zvi Assaraf and Orion 2005; Essex Report 2001; Hmelo-Silver and Pfeffer 2004; Sheehy et al. 2000) to investigate eleven- and twelve-year olds' development of systems thinking skills in the context of learning about a salt marsh utilizing a computer simulation game. This research study revealed that half of the participating students demonstrated three out of the seven skills on a pre-test, which were skills related to the identification of system elements, spatial boundaries of a system, and identification of factors causing certain patterns within the simulation. Students were successful at recognizing elements of a system but could not identify subsystems of a salt marsh according to post-test results. Despite the fact that subsystems were a part of the structure of the systems, it proved to be more difficult than identifying the isolated elements, primarily since students were unable to comprehend the relationship of the connected parts of the system. This finding supports Ben-Zvi Assaraf and Orion's (2005) claim that thinking skills are hierarchical and recognizes that understanding relationships is a higher-order skill that involves more than merely identifying the elements of a system. In this same study, Evagorou et al. (2009) found that elementary students displayed sophisticated systems thinking within this learning environment. Students showed improvement in both spatial and temporal boundaries and were able to infer the influence of change on a system.

The results reported by Evagorou et al. (2009) contrast with the findings of other studies which identify many of these skills as too difficult for younger students to attain (Ben-Zvi Assaraf and Orion 2005). Evagorou et al. (2009) reported that the learning environment of the simulation was effective in allowing students to engage with higher-level skills such as making predictions of effects on distant elements of a system or inferring the cause of a change from its consequences (Grotzer and Bell-Basca 2003b). Findings from this study and others question the hierarchical structure of developing system thinking skills, demonstrating that an appropriate learning environment can promote the development of complex systems thinking even at an early age (Hmelo-Silver and Pfeffer 2004). Although limited, research on the development of systems thinking with elementary students demonstrates that students at this age can develop systems thinking skills, possibly even more sophisticated skills, with appropriate instructional tools and a focus on systems thinking by the classroom teacher (Evagorou et al. 2009; Hmelo-Silver and Pfeffer 2004; Grotzer and Bell-Basca 2003b).

Recommendations for Using Representational Competence for Developing Systems Thinking in Classrooms

The implementation of systems thinking skills has often been neglected in the design of science learning environments (Golan and Reiser 2004). Aikenhead (2006) suggested that this neglect in instruction is a result of instruction that emphasizes facts and principles of science content instead of focusing on skills and thinking related to socio-humanistic perspectives of science. In most formal science educational settings, instruction is placed on the events that occur rather than studying how processes occur over time, on parts, and on isolated processes rather than demonstrating systemic relationships (Hannon and Ruth 2000). In many instances, students are left to figure out the connections and understanding of the interrelationships themselves, without appropriate scaffolding of the instruction (i.e. building on students' experiences and knowledge); students find thinking systemically to be difficult (Hmelo-Silver and Azevedo 2006). Although a limited amount of research is being conducted to investigate the appropriate learning strategies to develop systems thinking, this chapter provides some research evidence for designing and implementing instructional strategies using representational competence skills that have proven to be effective in science classrooms.

A variety of different instructional strategies using representations in science classrooms to develop systems thinking have been used with students ranging from elementary to high school (Ben-Zvi Assaraf and Orion 2005, 2010; Kali et al. 2003; Evagorou et al. 2009; Riess and Mischo 2010; Verhoeff et al. 2008; Liu and Hmelo-Silver 2009) including: computer modeling (e.g. simulations and hypermedia), authentic science context problems, multiple visual representations (e.g. photographs, diagrams, concept maps, and student drawings), hands-on experiences (e.g. indoor and outdoor inquiry-based labs), and knowledge integration activities (e.g. using representations).

Computer-based learning environments using modes of representations (e.g., simulations, hypermedia, modeling) were used successfully to promote the development of systems thinking while studying topics such as the ecosystem of a salt marsh (Evagorou et al. 2009), ecosystem forest (Riess and Mischo 2010), cellular biology (Verhoeff et al. 2008), and the human respiratory system (Liu and Hmelo-Silver 2009). Authentic science context, student scaffolding, multiple representations, and student reflections were implemented in combination in these studies, which led to a significant development of systems thinking. This combination of efforts is briefly discussed below as an effective classroom strategy to promote systems thinking and the development of representational competence.

The use of representations has been used in problem-based learning (a guided discovery approach of students working in groups to pose solutions for a presented problem) (Krajcik et al. 1998). In one case, Evagorou et al. (2009) challenged students to pose solutions for controlling mosquitoes for a village neighboring a salt marsh within a computer simulation. Students were provided with mosquito control as a real-life problem people encounter when living near a salt marsh. In another study that examined efforts to teach systems thinking, Liu and Hmelo-Silver (2009) had students investigate the human respiratory system by posing the question of "how do we breathe?" Verhoeff et al. (2008) had students examine the complexity of cellular biology by using a real-life context of breast-feeding. These authentic science contexts proved to be useful in providing students with real-life connections to the purpose of learning about systems.

The use of multiple representations has been associated with the development of systems thinking and representational competence (Treagust and Tsui 2013: Liu and Hmelo-Silver 2009). A hypermedia platform using multiple representations of information and visuals electronically linked in nonlinear format (i.e. displayed and linked as intricate web) has been proposed as effective in allowing multiple approaches of instruction (e.g., Moreno 2006). For example, Liu and Hmelo-Silver (2009) conducted an experiment to investigate two different organizational formats (i.e. function-centered and structure-centered representations) in a hypermedia format using the SBF conceptual representation framework of the human respiratory system and found students were able to identify and explain functions and behaviors of the hidden dimensions of this system using the function-centered format.

According to the SBF model (Hmelo et al. 2000; Hmelo-Silver and Pfeffer 2004) structures are defined as the elements of the systems, behaviors are known as the mechanisms of a system and, then finally, functions are the outcomes or roles of the system. Liu and Hmelo-Silver examined how these organizational formats impacted participants' understanding of salient (macro-level) phenomena involved in external respiration and non-salient (micro-level) phenomena involved in internal respiration. The researchers organized the content in the hypermedia in different formats using the SBF model. The function-centered format provided a more holistic understanding of the human respiratory system, beginning with the functions and behaviors at the top of a concept map, moving down to the structures oriented with those functions and behaviors. This organizational format directed and scaffolded students in the presentation of the information about this complex system. Students clicked on questions, such as, "Why do we breathe?" (function), which led students to pages with visuals and information answering the question explaining this particular function of the system. This function-centered format was designed to continually lead students through the process of going from functions to behaviors and finally, leading to the structures that make up the system. In contrast, the structure-centered format followed the traditional format much like a traditional textbook. The opening screen displayed a listing of the individual structures of the human respiratory system as isolated components of the system. The format directed students to click on an individual structure, which was linked to web pages explaining the isolated structure and how the structure connects to their respective behaviors and functions. The content information in both formats was identical, except for the organization format of the presented information. The ability to distinguish between these hidden dimensions of the system (micro-level) and the ability to distinguish between the function and behaviors of systems were barriers for developing systems thinking. Positive results were found in the use of the function-centered format on participants' understanding of the non-salient (micro-level) phenomena of the internal respiratory system. These findings also contribute to our understanding of the role of students' abilities in developing representational competence by comparing and contrasting various representations and their content information and being able to make connections across different representations by explaining the relationships between them (Kozma and Russell 2005).

This SBF conceptual representation model could be applied in science instruction more broadly to help students understand other kinds of complex systems and representational competence by organizing the presented information in a function-centered structure, beginning with the functions and behaviors of a system, and then leading to the structures of the system. The selection of representations and how they are presented can play a critical role in scaffolding students' understandings of a system.

The use of multiple visual representations has been shown to help people learn complex new ideas (Ainsworth 2006). The process of modeling can take students through an iterative process of formation, revision, and elaboration of models that compare each model while assisting students in gaining a deeper understanding of the system. Verhoeff et al. (2008) used a systems approach of modeling for developing an understanding of cellular biology. The modeling process utilized multiple visual representations and physical models. The systems approach required students to engage in thinking back and forth between computer-constructed models, real cells (observed through an electron microscope), student-created physical models, and visual representations, such as diagrams found on the Internet. In each phase of the process, students were required to engage in discourse, regarding the process of breast-feeding at the cellular level using the multiple models to explain the functions of each of the structures found in the cells, which led to milk production. This modeling process proved to be a powerful visualization tool, which led students to understand both the dynamics of (cell) biological processes and the hierarchical structure of biological systems. Secondary science teachers could implement this systems approach to modeling by engaging students in the thinking back-and-forth strategy while reflecting on the visual representations, which differ in abstractness, with the inclusion of the biological objects themselves. While engaged in the modeling process, teachers can develop representational competence by having students make predictions, draw inferences and make connections among representations.

The use of multiple visual representations and the use of systems modeling with the explicit use of teacher scaffolding were shown to be effective instructional strategies for promoting systems thinking in science (Liu and Hmelo-Silver 2009; Evagorou et al. 2009; Verhoeff et al. 2008; Riess and Mischo 2010). Liu and Hmelo-Silver (2009) also emphasized the importance of teacher guidance in helping students focus on the science principles associated with the systems and making connections while exploring the hypermedia format. When using computer simulations of a salt marsh with elementary students, Evagorou et al. (2009) found that students who did not receive instruction or scaffolding from their teacher did not grow significantly in their development of systems. Another study by Riess and Mischo (2010) also reported that teachers were critical in developing representational competence in systems contexts. Riess and Mischo (2010) used computer simulations to declare the most effective method (i.e. computer simulation, specific lessons, or a combination of both) to teach about ecosystems. Students were randomly assigned to a treatment group for their assigned learning environment. The study confirmed the effectiveness of computer simulations when teacher guidance and scaffolding was provided to promote systems thinking and representational competence.

Computer simulations as effective representation tools allow students to manipulate and explore complex systems that may be too large or too small for direct observation. One of the challenges encountered by teachers is providing access to a system to collect data without having to devote excessive time and repeated field-study visits (NRC 2000). Interactive simulations enable teachers to bypass this challenge (Evagorou et al. 2009). Students are able to have continuous access to the system, allowing them to explore the various parameters of the systems' structures, functions, and behaviors.

Hands-on experiences (outside and inside labs) in combination with knowledge integration activities using representations can be used to illustrate the growth of systems thinking development. As discussed previously, Ben-Zvi Assaraf and Orion (2005) developed a STH model built on eight characteristics of a sequential growth of levels in systems thinking. In studies of elementary and middle school students, there were difficulties in reaching the implementation level (i.e. highest level), which included skills such as making generalizations in regards to the system, describing the processes that occur in the hidden dimensions, and explanations of the relationship to time (i.e. prediction and retrospect). Based on these studies, Ben-Zvi Assaraf and Orion (2010) suggested that elementary teachers begin instruction of systems at the analysis level (i.e. identify processes and structures) and then move on to the first two parts of the synthesis level (i.e. identifying relationships between two components and identifying dynamic relationships within the system). They argued that if elementary teachers can provide this foundation for systems thinking, then middle school teachers can focus their instruction on the implementation level. This hierarchical structure still needs to be evaluated and used with more students and within more studies of various complex systems to be empirically sound. Although its hierarchical structures need more scrutiny with research, the levels themselves provide a solid framework for designing systems curriculum in classrooms. Teachers can implement learning experiences such as labs, field experiences, and integration activities using representations to begin building this knowledge of systems thinking.

Even though the research is limited in the use of systems thinking strategies in classrooms, there are some common recommendations that science educators and teachers can utilize. The use of inquiry-based labs connecting indoor labs and outdoor field experiences, followed by knowledge integration activities, has been shown to be an essential component for developing systems thinking. Science educators and teachers are familiar with these methods; the main impetus of change should focus on how students think in terms of a system and the process of evaluating knowledge through the use of representations. The use of multiple visual representations and computer modeling have also shown to be effective methods for promoting systems thinking; again, scaffolding is a key component of instruction to assist students in developing the skills necessary to make sufficient growth. This scaffolding can be implemented in various ways: through probing specific questions, engaging in dialog, and allowing time for knowledge integration activities that provide time for reflection of the functions and behaviors of the systems while explicitly using representations.

Teachers' Selection of Types of Representations

There is a limited amount of research documenting the reasons why elementary teachers select specific representations when teaching about a complex system. The ability to select and explain a scientific concept found in a representation is one aspect of representational competence. The use of technology and the lack of science textbooks in elementary classrooms have placed elementary teachers in the position of creating their own instructional materials. In a recent study investigating the processes taken by elementary in-service teachers (n = 67) and elementary pre-service teachers (n = 69) regarding their selection of representations when planning a lesson on a complex system (e.g., water cycle) (Lee and Jones 2017) a rubric was created for categorizing teachers' selection reasons.

For this study, elementary in-service and pre-service teachers completed a card sort task in which they selected pictorial representations (e.g. diagrams, photographs, and maps) to teach about the water cycle. A panel of experts selected 15 pictorial representations from the Internet for the card sort task to use in a lesson about the water cycle. The elementary in-service and pre-service teachers described why they selected the representation. A rubric was developed to analyze reasons for selecting representations. The rubric emerged from a content analysis that began with the researcher reading and re-reading the data to formulate a tentative understanding of the data (Roth 1995). Several iterations of content analysis were conducted to establish nine themes for coding of selection and non-selection rationale. A science educator and researcher separately coded a randomly selected 10% of data. As part of the verification methodology (Strauss 1987), the coding was repeated three times (coding, discussion, revision) until inter-rater reliability was reached at 97%.

The four themes that emerged from the coding are included in Table 1. The indicated thematic categories are not exclusive: a given response may fit in more than one category.

The rubric presented here can be used as a tool to help teachers recognize pedagogical preferences for selecting images to teach about complex systems. Two examples of representations were chosen from the study to illustrate the use of this coding rubric. The first example of a representation is a digital diagram titled *The Water Cycle* from the U.S. Geological Survey (USGS 2014) website. The diagram illustrates the entire system of the water cycle, identifying the components and processes of the system, and shows the hidden dimensions of groundwater as well as

Table 1 Coding rubric for teachers' selection reasons for representations

Thematic categories	Description	Teachers' selection rationale
+ Aesthetics	Responses referring to color, brightness; refers to the look or appearance of the pictorial representation.	*Colorful* *Pleasing in appearance* *Attractive* *Clear graphic* *Great visual*
− Non-aesthetics	Responses refer to unattractive aspects of the representation.	*Ugly picture* *Not appealing* *Plain not attractive*
+Understandable	Responses refer to understandability, simple student accessibility, or developmentally appropriate.	*Kid friendly example* *Visual is easy to understand* *Good vocabulary* *Simple and concise* *Lower grade students would* *Be able to understand it.*
−Complexity	Responses refer to elements that may cause confusion (arrows, lack of labels, difficult terminology), are confusing, or are hard to understand.	*Too many arrows may lead to confusion* *Photos aren't clear as to what they are showing me so I doubt an elementary student would understand* *I don't understand it*
+ Systems thinking	Responses refer to representations of a component or process of the water cycle system, or identify interactions within the system.	*Image gives a good description of mountain & runoff that goes into the ocean.* *It demonstrates how water system changes overtime.* *Shows water as a cycle* *Shows runoff* *Illustrates precipitation*
− Non-systems thinking	Responses refer to a lack of conceptual connection between the representation and the water cycle.	*This image I think would be better for the topic of sewage and pollution.* *Not sure how to use it when talking about the water cycle.* *I don't understand how it relates to the water cycle.*
+ Relevance	Responses refer to relevancy to students' lives or as an illustration of real life examples.	*It is detailed realistic* *Real example* *Student will relate to this picture something they see everyday*
−Non-relevance	Responses refer to the representation as not relevant or as uninteresting to students.	*The students wouldn't be interested in this map.* *Runoff into a stream isn't a picture that really needs to be seen.* *Not related to the curriculum.*

Note: Selection categories and reasons signified by + symbol; non-selection categories and reasons signified by − symbol

processes such as transpiration. Teachers identified the following aesthetic and understandable reasons for selection of this diagram: "bright colorful basic," "clearly labeled and easy to read," and "colorful pleasing-to-eye informative." The non-selection reasons of "non-aesthetic," "complexity," and "non-systems think-ing" suggest teachers' limited representational competence when selecting a dia-gram for classroom instruction about a complex system: "not very clear about what is what," "seems a little complicated," "too many arrows pointing different ways," and "unclear where main points of cycle are condensation, evaporation, etc." These statements regarding selection and non-selection reasons provide clear differences in how teachers view the same representation. Teachers tended to select diagrams based on aesthetic and understandable reasons. The understandable reasons were related to the representation being easy to read or understand and being develop-mentally appropriate for students. These selection or non-selection aesthetics rea-sons by teachers suggest that the selection process is not necessarily about the science content being presented in the diagram but more about the appearance. This reason for selection suggests that teachers may not be considering the development of representational competence in their selection process. Not using representa-tional competence in their selection process also suggests teachers may not consider the representation as a communicative tool of science. The absent of representa-tional competence should be addressed in teacher preparation programs as well as professional development for in-service teachers.

The second representation was a photograph from the Bird Education Network website (2010) of a bird that was involved in an oil spill in the ocean. The statements of teachers regarding the selection of this photograph revealed information about their selection process, but more importantly, also revealed information about their application of systems thinking. Teachers reported selecting the image because it showed "what can happen when oil is spilled or dumped in water," "what polluted water can affect," and "why we need to take care of our water." These selection reasons expose teachers' understanding of the human impact on the water cycle. The selection rationales also suggest teachers' pedagogical applications for this rep-resentation such as the photograph to lead students in a discussion of the relation-ships between pollution, animals, and humans, within the water cycle system. The non-selection reasons also offered information about teachers' application and knowledge of systems thinking as well as their representational competence. Here are a few examples of the non-systems thinking as identified by the rubric: "what's the purpose for this? Students may wonder the same thing as me, it is more about human pollution rather than the water cycle," "nothing to do with the water cycle," "an oil spill victim has little to do with the basic water cycle" and "not sure how to integrate this image." These non-systems thinking statements by teachers suggest limited representational competence in understanding how to interpret this photo-graph and its relation to the water cycle system and applying systems thinking. Teachers' can use the developed rubric as a way to critique their selection of repre-sentations when teaching about systems (Fig. 1).

These two examples illustrate the need to understand more about why teachers select and use certain representations to teach about systems in science. Teachers

a
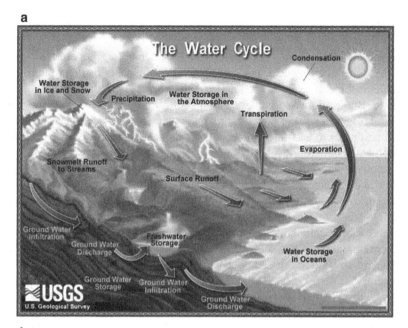

b

Fig. 1 Examples of pictorial representations: (**a**) The U.S. Geological Survey (2014). U.S. Department of the Interior, U.S. Geological Survey (**b**) Bird Education Network (2010). Flying WILD activity teaches how oil in water effects birds

need to be able to scaffold and support students' use of the representations and provide experiences with multiple representations into instruction in order to help communicate the dynamics of systems and to promote systems thinking.

Conclusions

The recent call for teaching systems thinking challenges teachers to find ways to make the dynamic, complex, and less obvious components of the system understandable to students. Representations provide teachers with the tools to meet this challenge. But as discussed here, representations are imprecise models of systems and teachers need a level of representational competence to select images that move beyond aesthetics to accurately represent components of systems in ways that are developmentally appropriate for students.

Strategies that teachers can use to teach systems thinking include using authentic contexts and inquiry-based investigations in combination with representations. Keys to making connections to systems thinking include using models and modeling both as teacher pedagogical tools for whole class instruction and as a means for students to individually build representations as a part of the meaning making process. Representations allow students to search for components and processes, interrelationships between these components and processes, and hidden dimensions of the system.

Finally, this chapter offers some classroom recommendations for teachers such as establishing an authentic science context (problem) for students to investigate in regards to the system of study; using indoor and outdoor inquiry-based labs; connecting the experiences with knowledge integration activities; using multiple visual representations and modeling; and, of course, scaffolding students' progress as they work to develop the skills of systems thinking and representational competence. These recommendations are not new to science educators or science teachers; the key difference in implementing these strategies is the explicit approach to systems thinking and the use of representations. The systems thinking instructional approaches should address the search for patterns in systems, visualize interactions among the system components, find cause and effect relationships, and conceptualize the hidden dimensions of a system.

Teachers and students have access to a plethora of representations through the Internet in unprecedented volumes. Science teacher educators have a new challenge to teach teachers how to select and use representations to teach accurate science content while modeling complex systems phenomena. Perhaps equally important is teaching students to be critical consumers and to use representations as learning tools.

References

Aikenhead, G. S. (2006). *Science education for everyday life: Evidence-based practice*. New York: Teachers College Press.

Ainsworth, S. (2006). DeFT: A conceptual framework for considering learning with multiple representations. *Learning and Instruction, 16*, 183–198.

Ben-Zvi Assaraf, O., & Orion, N. (2004). *Learning about Earth as a system: A new approach of designing environmental curricula*. Retrieved from http://www.weizmann.ac.il/g-earth.

Ben-Zvi Assaraf, O., & Orion, N. (2005). Development of system thinking skill in the context of earth system education. *Journal of Research in Science Teaching, 42*, 518–560.

Ben-Zvi Assaraf, O., & Orion, N. (2010). System thinking skills at the elementary school level. *Journal of Research in Science Teaching, 47*, 540–563.

Bird Education Network. (2010). Flying WILD activity teaches how oil in water effects birds. [online image] *Council for Environmental Education*, Retrieved from http://www.birdeducation.org/BENBulletin32.htm.

Booth Sweeney, L. (2000). *Bathtub dynamics: Initial results of a systems thinking inventory.* Retrieved from http://web.mit.edu.jsterman/www/bathtub.pdf.

Booth Sweeney, L., & Sterman, J. D. (2007). Thinking about systems: Student and teacher conceptions of natural and social systems. *System Dynamics Review, 23*, 285–312.

Dorner, D. (1980). *The logic of failure.* New York: Metropolitan Books/Henry Holt.

Draper, F. (1993). A proposed sequence for development system thinking in a grades 4–12 curriculum. *System Dynamic Review, 9*, 207–214.

Emery, R. E. (1992). Parenting in context: Systemic thinking about parental conflict and its influence on children. *Journal of Consulting and Clinical Psychology, 60*, 909–912.

Essex Report. (2001). The future of system dynamics and learner-centered learning in K-12 education. Paper presented at the *International System Dynamics Society Conference*, Essex, MA.

Evagorou, M., Korfiatis, K., Nicolaou, C., & Constantinou, C. (2009). An investigation of the potential of interactive simulations for developing system thinking skills in elementary school: A case study with fifth-graders and sixth-graders. *International Journal of Science Education, 31*, 655–674.

Faughman, J. G., & Elson, R. (1998). Information technology and the clinical curriculum: Some predictions and their implication for the class of 2003. *Academic Medicine, 73*, 766–769.

Fordyce, D. (1988). The development of systems thinking in engineering education: An interdisciplinary model. *European Journal of Engineering Education, 13*, 283–292.

Forrester, J. (2007). System dynamics-a personal view of the first fifty years. *System Dynamics Review, 23*, 245–358.

Frank, M. (2000). Engineering systems thinking and systems thinking. *Systems Engineering, 3*, 63–168.

Golan, R., & Reiser, B. (2004). Investigating students' reasoning about the complexity manifested in molecular genetics phenomena. Proceedings from *American Educational Research Association*, San Diego, CA.

Goldstone, R. L., & Wilensky, U. (2008). Promoting transfer complex systems principles. *Journal of the Learning Sciences, 17*, 465–516.

Graczyk, S. L. (1993). Get with the system: General systems theory for business officials. *School Business Affairs, 59*, 16–20.

Grotzer, T., & Bell-Basca B. (2003a). *Helping students to grasp the underlying causal structures when learning about ecosystems: how does it impact understanding?* Paper presented at the *National Association for Research in Science Teaching Annual Conference*, Atlanta, GA.

Grotzer, T. A., & Bell-Basca, B. (2003b). How does grasping the underlying causal structures of ecosystems impact students' understanding? *Journal of Biological Education, 38*(2), 16–30.

Hannon, B., & Ruth, M. (2000). *Dynamic modeling* (2nd ed.). New York: Springer.

Hmelo, C. E., Holton, D., & Kolodner, J. L. (2000). Designing to learn about complex systems. *Journal of the Learning Sciences, 9*, 247–298.

Hmelo-Silver, C. E., & Azevedo, R. A. (2006). Understanding complex systems: Some core challenges. *Journal of the Learning Sciences, 15*, 53–61.

Hmelo-Silver, C. E., & Pfeffer, M. G. (2004). Comparing expert and novice understanding of complex system from the perspective of structures, behaviors, and functions. *Cognitive Science, 28*, 127–138.

Hmelo-Silver, C. E., Marathe, S., & Liu, L. (2007). Fish, swim, rocks sit, and lungs breathe: Expert-novice understanding of complex systems. *The Journal of The Learning Science, 16*, 307–331.

Jacobson, M., & Wilensky, U. (2006). Complex systems in education: Scientific and educational importance and implications for the learning sciences. *Journal of the Learning Science, 15*, 11–34.

Kali, Y., Orion, N., & Eylon, B. (2003). Effect of knowledge integration activities on students' perception of the earth's crust as a cyclic system. *Journal of Research in Science Teaching, 40*, 545–565.

Kim, D.H. (1999a). Introduction to system thinking. In: System thinking tools and applications. ASA: Pegasus Communications.

Kim, J. (1999b). Making sense of emergence. *Philosophical Studies, 96*(2–3), 3–36.

Klahr, D., & Simon, H. A. (1999). Studies of scientific discovery: Complementary approaches and convergent findings. *Psychological Science, 15*, 661–667.

Kozma, R., & Russell, J. (1997). Multimedia and understanding: Expert and novice responses to different representations of chemical phenomena. *Journal of Research in Science Teaching, 34*, 949–968.

Kozma, R., & Russell, J. (2005). Students becoming chemists: Developing representational competence. In J. K. Gilbert (Ed.), *Visualizations in science education* (pp. 121–146). Dordrecht: Springer.

Kozma, R., Chin, E., Russell, J., & Marx, N. (2000). The roles of representations and tools in the chemistry laboratory and their implications for chemistry learning. *Journal of the Learning Sciences, 9*, 105–143.

Krajcik, J., & Sutherland, L. M. (2010). Supporting students in developing literacy in science. *Science, 328*, 456–459. https://doi.org/10.1126/science.118293.

Krajcik, J., Blumenfeld, P. C., Marx, R., Bass, K. M., Fredrick, J., & Soloway, E. (1998). Inquiry in project-based science classrooms: Initial attempts by middle school students. *Journal of the Learning Science, 3*, 313–350.

Kress, G., Jewitt, C., Ogborn, J., & Tsatsarelis, C. (2001). *Multimodal teaching and learning: Rhetorics of the science classroom*. London: Continuum.

Lee, T., & Jones, M. G. (2017). Elementary Teachers' Selection and Use of Visual Models. *Journal of Science Education and Technology., 27*(1), 1–29.

Lemke, J. L. (2004). The literacies of science. In E. W. Saul (Ed.), *Crossing borders in literacy and science instruction: Perspectives on theory and practice* (pp. 33–47). Arlington: International Reading Association.

Lewis, J. P. (1998). *Mastering Project Management: Applying advanced concepts of systems thinking, control, and evaluation, resource allocation*. New York: McGraw Hill.

Liu, L., & Hmelo-Silver, C. E. (2009). Promoting complex systems learning through the use of conceptual representations in hypermedia. *Journal of Research in Science Teaching, 46*, 1023–1040.

Mandinach, E. B. (1989). Model-building and the use of computer simulation of dynamic systems. *Journal of Educational Computing Research, 5*, 221–243.

Martin, B. L., Mintzes, J. J., & Clavijo, I. E. (2000). Restructuring knowledge in biology: Cognitive processes and metacognitive reflections. *International Journal of Science Education, 22*, 303–323.

Mason, C. L. (1992). Concept mapping: A tool to develop reflective science instruction. *Science Education, 76*, 51–63.

Moreno, R. (2006). Learning with high tech and multimedia environments. *Current Directions in Psychological Science, 15*, 63–67.

National Research Council. (2000). How people learn: Brian, mind, experience and school. In J. D. Bransford, A. L. Brown, R. R. Cocking, & S. Donovan (Eds.), *Committee on developments in the science of learning and committee on learning research and educational practice*. Washington, D.C.: National Academy Press.

National Research Council. (2007). *Taking science to school: Learning and teaching science in grades K-8*. Washington, D.C.: The National Academies Press.

National Research Council. (2012). *A framework for K-12 science education: Practices, crosscutting concepts, and Core ideas*. Washington, D.C.: The National Academies Press.

NGSS Lead States. (2013). *Next generation science standards: For states, by states*. Washington, D.C.: The National Academies Press.

Nitz, S., Ainsworth, S., Nerdel, C., & Prechtl, H. (2014). Do students perceptions of teaching predict the development of representational competence and biological knowledge? *Learning and Instruction, 31*, 13–22.

Norris, S., & Phillips, L. (2003). How literacy in its fundamental sense is central to scientific literacy. *Science Education, 87*(2), 224–240.

Novak, J., & Gowin, D. B. (1984). *Learning how to learn*. Cambridge, UK: Cambridge University Press.

Ossimitz, G. (2000). Teaching system dynamics and systems thinking in Austria and Germany. Proceedings from the *18th International Conference of System Dynamics Society*, Bergen, Norway.

Penner, D. (2000). Explaining systems: Investigating middle school students' understanding of emergent phenomena. *Journal of Research in Science Teaching, 37*, 784–806.

Resnick, L. B. (1987). *Education and learning to think*. Washington, D.C.: National Academy Press.

Riess, W., & Mischo, C. (2010). Promoting systems thinking through biology lessons. *International Journal of Science Education, 32*, 705–725.

Roth, W. M. (1994). Students' views of collaborative concept mapping: An emancipator research project. *Science Education, 78*, 1–34.

Roth, W. M. (1995). Affordances of computers in teacher-student interactions: The case of interactive PhysicsTM. *Journal of Research in Science Teaching, 32*, 329–347.

Sabelli, N. H. (2006). Complexity, technology, science, and education. *Journal of the Learning Sciences, 15*, 5–9.

Senge, P. M. (1990). *Fifth discipline: The art and practice of the learning organization*. New York: Doubleday.

Sheehy, N. P., Wylie, J. W., McGuinness, C., & Orchard, G. (2000). How children solve environmental problems: Using computer simulations to investigate system thinking. *Environmental Education Research, 6*, 109–126.

Steed, M. (1992). Stella, a simulation construction kit: Cognitive processes and educational implications. *Journal of Computers in Mathematics and Science, 11*, 39–52.

Stieff, M. (2011). Improving representational competence using molecular simulations embedded in inquiry activities. *Journal of Research in Science Teaching, 48*, 1137–1158.

Strauss, A. L. (1987). *Qualitative analysis for social scientists*. New York: Cambridge University Press.

Treagust, D. F., & Tsui, C. Y. (Eds.). (2013). *Multiple representations in biological education, models and modeling in science education 7*. Dordrecht: Springer.

U.S. Geological Survey. (2014). *The Water Cycle* [digital image]. Retrieved from http://water.usgs.gov/edu/watercycle-screen.html.

Ullmer, E. J. (1986). Work design in organizations: Comparing the organizational elements models and the ideal system approach. *Educational Technology, 26*, 543–568.

Verhoeff, R. P., Waarlo, A. J., & Boersma, K. T. (2008). Systems modeling and the development of coherent understanding of cell biology. *International Journal of Science Education, 30*, 331–351.

Wilensky, U., & Resnick, M. (1999). Thinking in levels: A dynamic systems approach to making sense of the world. *Journal of Science Education and Technology, 8*, 3–19.

Yore, L. D., & Hand, B. (2010). Epilogue: Plotting a research agenda for multiple representations, multiple modality, and multimodal representational competency. *Research in Science Education, 40*, 93–101.

Yore, L. D., Pimm, D., & Tuan, H.-L. (2007). The literacy component of mathematical and scientific literacy. *International Journal of Science and Mathematics Education, 5*, 559–589.

Leveraging on Assessment of Representational Competence to Improve Instruction with External Representations

Mounir R. Saleh and Kristy L. Daniel

Introduction

Scientists' progress in understanding complex natural phenomena dictated modern textbooks that are charged with external representations to enhance text comprehension. Approximately, one-third of page space is occupied by images in recent editions of representative science textbooks (Griffard 2013). In spite of their pedagogical potentials, students' abilities to interpret and use these representations is often overlooked (Kozma and Russell 2005). In this chapter, we highlight the importance of assessing these abilities through utilizing valid and reliable assessment practices. Additionally, we discuss some of the different views that researchers have on the challenges that face students while learning/problem-solving with external representations. We show here how each of these views defines potential challenges and accordingly provides a pool of treatment protocols. The described assessment practices, in turn, help diagnose which challenge(s) are hindering the learning process and hence assist in choosing the appropriate treatment.

Some of these views belong to the cognitive load theory (CLT) (Sweller and Chandler 1994), the cognitive theory of multimedia learning (CTML) (Mayer 2009), and the multiple external representations (MER) framework (Treagust and Tsui 2013). In CLT, an external representation is considered as challenging/complex if it consists of six or more interacting elements (Sweller and Chandler 1994). From CLT researchers' standpoint, for a novice learner to locate a point $P(x, y)$ on a coordinate system, they need to simultaneously consider the following seven elements:

M. R. Saleh (✉)
Bahrain Teachers College, University of Bahrain, Zallaq, Kingdom of Bahrain

K. L. Daniel
Texas State University, San Marcos, TX, USA
e-mail: kristydaniel@txstate.edu

© Springer International Publishing AG, part of Springer Nature 2018
K. L. Daniel (ed.), *Towards a Framework for Representational Competence in Science Education*, Models and Modeling in Science Education 11,
https://doi.org/10.1007/978-3-319-89945-9_8

1. The x axis is a graduated, horizontal line; the y axis is a graduated, vertical line. These two lines cross at the zero point on both axes, called the origin, and are at right angles because one is vertical and the other horizontal.
2. P in P(x, y) refers to the relevant point in both the algebraic and geometric systems.
3. x in P(x, y) refers to a location x on the x axis.
4. y in P(x, y) refers to a location y on the y axis.
5. Draw a line from x on the x axis at right angles to the axis.
6. Draw a line from y on the y axis at right angles to the axis.
7. The point where these two lines meet is $P(x, y)$. (Sweller and Chandler 1994, pp.190–191)

The driving assumption here is that these elements cannot be learned in isolation (Pollock et al. 2002). Processing them at once is then likely to overload the novice's working memory and hamper the learning process. On the other hand, CLT researchers acknowledge that an advanced learner can locate $P(x, y)$ with minimal consciousness. Accordingly, they perceive this source of complexity as a function of expertise (Kalyuga 2007). For them, this challenge can be overcome through practice, which enables the learner to process these elements automatically rather than consciously in the working memory (Pollock et al. 2002). Several CLT treatment protocols were devised for this purpose, such as the sequencing techniques described in Pollock et al. (2002), Van Merriënboer et al. (2003), Van Merriënboer et al. (2006), and Van Merriënboer and Sweller (2005).

CTML researchers seem to agree with CLT's definition of the potential challenges imposed by external representations. In CTML terms, complexity may arise from the need to understand presented "material that consists of many steps and underlying processes" (Mayer 2009, p. 80). A review of the introduced concepts in CTML lessons reveals explanations of cause-and-effect chains of events (e.g. car braking system, air pumps) (Mayer 2009). Hence, the "steps" mentioned in the definition are also naturally interacting. One CTML treatment to complexity is the pretraining technique in which the learner attends to major elements in a representation before attending to their interactivity (Mayer 2009). Another treatment, the modality technique, utilizes external representations along with spoken rather than written words. This approach builds on Paivio's dual-coding assumption that people possess separate channels for processing visual and auditory information (Paivio 1986). According to this theory, presenting some information elements in spoken words permits their processing in the auditory channel thus unloading the visual channel. Referential connections between the two channels would then help better process the presented material. A third CTML treatment is the signaling technique in which the essential organization of a representation is highlighted through arrows, distinctive colors, pointing gestures, etc. (Mayer 2009). Still when captions are used along with representations, CTML researchers recommend presenting words near their corresponding representation elements to avoid holding several elements in working memory while searching for relevant connections. This sample of treatment protocols shows how CTML researchers deal with the challenges that might be imposed by their own external representations, mainly words and animations. Yet, sometimes students face instructional material that is represented via multiple representations rather than just two representations. The MER framework sufficed this need.

According to MER framework, the challenge in some external representations may arise from the need to engage in abstract thinking necessary to connect information elements at different levels of organization in the studied system (e.g. macro, submicro, symbolic, etc.) (Treagust and Tsui 2013). This issue is highly likely in disciplines such as biology and biochemistry where the learner has to move back and forth from the symbolic level (e.g. what is the mechanism of glycolysis) to, the submicro level (e.g. which enzymes and other molecules are involved), the micro level (e.g. where in the cell glycolysis is taking place), and the macro level (e.g. how does glycolysis relate to the entire energy profile of tissues within organs, systems, organisms, etc.). Linking between these levels is likely to overload the learner's working memory as it "often requires bridging across a great cognitive distance" (Schönborn and Bögeholz 2013, p. 122). According to this framework, this sort of challenge can be met by using multiple external representations that would support the learner in shifting between the different levels (Schönborn and Bögeholz 2013). According to this framework, the support that MERs offer to the learner stems from the notions that: "(1) specific information can best be conveyed in a particular representation, [and] (2) several representations can be more useful in displaying a variety of information" (Treagust and Tsui 2014, p. 312).

To summarize, each of the three views adds something to our understanding of the challenges that may face students while learning from external representations. CLT offers the element interactivity explanation, CTML builds on CLT's definition along with the dual-channel assumption, and MER highlights the need for hierarchical translations in certain domains of knowledge. Hence, it is beneficial to consider the three perspectives when studying students' representational competencies. The following sample problem from CTML research emphasizes this need (Mayer et al. 2002). In this experiment, the learner is taught that the car braking system works as such:

> The car's brake is pushed down, which causes a piston to move forward in the master cylinder, which causes the fluid to become compressed in the tube, which causes the smaller pistons to move forward in the wheel cylinders, which causes the brake shoe to move forward, which causes the brake drum to be pressed against, which causes the wheel to slow down or stop. (Mayer et al. 2002, p. 147)

In a subsequent problem, the learner may be asked about the consequences of replacing the brake shoe with one that has a different (say smaller) surface contact with the brake drum. Based on the provided external representation, the learner may easily modify the causal model described above to be as such, "the car's brake is pushed down ... which causes the brake shoe to move forward, which causes the brake drum to be pressed against [less efficiently], which [will take longer for] the wheel to slow down or stop." Mind here that the answer is based on a surface feature of the representation, which is the degree of contact between the brake shoe and the

drum. Nevertheless, if in another problem the learner were asked about the consequences of replacing the brake shoe by another with a different contact material
with the brake drum, then solving the problem would require selecting relevant
frictional properties of the given contact material (such as ability to recover quickly
from increased temperature). Notice in this case that the answer requires deeper
understanding of the underlying principle that stops the wheel (i.e. the friction
between the shoe and the drum which converts kinetic energy of the car to thermal
energy). Another challenge in this problem is the need to cognitively move back and
forth along the hierarchical levels of organization in the braking system from the
symbolic level (e.g. switching between forms of energy), to the submicro level (e.g.
molecules constituting the contact material), to the macro level (e.g. bringing the
wheel to a stop).

Therefore, in moving from the first to the second problem, the learner may display different levels of representational competence based on: (1) their level of
understanding of the underlying principle (e.g. friction), (2) degree of automation of
recalled facts while solving the second problem (e.g. frictional properties of the
given contact materials), and (3) ability to mentally shift between the levels of organization in the presented system (e.g. symbolic, submicro, and macro levels).

Collectively, the three perspectives provide treatments for each of these cases:
(1) CTML pre-training can be employed to foster understanding of the underlying
principle (Mayer 2009), (2) automation of facts can be achieved through utilizing
CLT's part-whole task sequencing technique (Van Merriënboer et al. 2006), and (3)
shifting between the various levels of the studied system can also be supported by
multiple representations (MERs) of these levels (Schönborn and Bögeholz 2013).
With these perspectives offering explanation and treatment, the three have the
potential to constitute an emerging framework for assessing representational competence. Yet, this should be accompanied with traditional assessment practices
because, for instance, one cannot tell which of the three treatments is to be employed
unless highlighted with Item analysis. Additionally, some conceptual/methodologi-
cal issues call for concerns. Below, we discuss these concerns and how they can be
overcome with traditional assessment measures.

The first concern is with CLT's notion of element interactivity, which does not
seem enough to explain what constitutes a complex representation. For instance,
Klahr and Robinson (1981) reported that children experienced more difficulty solving nontraditional Tower of Hanoi problems although the required number of moves
as well as the number of pegs and disks (interacting elements) is the same. In cases
as such, adopting the CLT perspective does no help in explaining differences in
representational competence among students. The Tower of Hanoi problem hints to
the fact that the degree to which a student can interpret a representation is also influenced by the range of cognitive processes required to answer the questions that are
based on this representation. Some basic questions assess students' retention of the
presented material. Other questions however build on this retention to assess higher
levels of intellectual behavior such as *"Applying"* and *"Analyzing."* Therefore, an
interesting approach to this issue is to gauge the change in representational competence through utilizing assessment instruments that consist of a continuum of

problems based on the hierarchical cognitive orders in revised Bloom's taxonomy (Anderson et al. 2001). Potential benefits of such an approach will be discussed later in this chapter.

Another concern is about the external validity of CTML's research-based treatments. A substantial portion of participants in CTML research are psychology students who are asked to learn about topics beyond their area of expertise (de Jong 2010). Examples of these learning topics included external representations on how airplanes achieve lift (Mautone and Mayer 2001) and other topics of special interest to science students (e.g. Madrid et al. 2009; Mayer et al. 2002; Moreno 2004; and Stull and Mayer 2007). The concern here is whether non-science students can interpret these representations much like their science counterparts, given the difference in prior knowledge. In the airplane experiment for instance, a non-science student may not be competent enough to understand from the representation why the air flowing on top of the wing is faster than that under the wing, even though they are instructed that it has to travel a curved (longer) surface in the same amount of time. This potential problem arises from the fact that non-science students may not recall the mathematical relationship between speed, distance, and time. It is true that this can be overcome through applying CTML pre-training technique in which the learners' memory is refreshed about this mathematical relationship before instruction takes place (Mayer 2009). However, common assessment practices by CTML researchers do not have the power to diagnose this potential problem in the first place as they solely rely on counting the number of correct responses on a test (see a list of 90 plus experiments on Appendix B in Dacosta 2008). Yet, as we will demonstrate, employment of assessment instruments that discriminate among learners of different prior knowledge can resolve these issues. This becomes even handier when supported with some Item analysis such as distractor[1] analysis.

The concern with MER framework is that explicit representation of the various levels of organization necessitates building proper referential connections among the given external representations. This even becomes more challenging when these representations are provided in different modes (e.g. animations, graphs, equations, words, etc.). Kozma and Russell (2005) for instance found that novices tend to connect representations to one another based on their media type (e.g. all graphs, all equations) rather than their complementary function to explain a concept. Therefore, students' differences in translation skills along the different modes of representation would be a limiting factor for the success of this strategy (Kozma and Russell 1997). To overcome this limitation, Kozma and Russell (1997) called for adopting a curriculum with assessment practices that evaluate these translation skills. This measure would then force teachers to teach these skills hence bridging gaps among learners.

To wrap up, we considered element interactivity as a potential challenge for learning with representations. We also touched on cases where representational competence is influenced either by, cognitive orders of assigned problems, learner's prior knowledge, or the degree of translation skills needed to navigate between

---

[1] A distractor is a multiple choice answer meant to distract a test taker from the correct answer.

different levels of organization/modes of representation. In the following sections, we provide an example for how assessment instruments may be designed to diagnose each case. However, to consider claims stemming from data generated by these instruments, the data itself should first be validated. Accordingly, we display some holistic and Item psychometric analyses that we recommend for the success of such assessment practices. We do this by first describing how the content of the given instrument is defined as challenging in CLT, CTML, and MER's terms. We then establish validity checks including content, face, and discriminant validities. These validity checks comprise the entire continuum of problems in the instrument, which is based on the cognitive orders in revised Bloom's taxonomy. Additionally, we do Item analysis to remove poorly performing Items. The implications of these different analyses follow in the discussion section.

Instrument Content

As for the knowledge dimension, the presented instrument targets students' conceptual knowledge of "enzyme specificity," which is a core concept in college science education. Our choice of this knowledge dimension originates partly from the fact that many of the often used multimedia lessons in CTML research aim at explaining *concepts* more often than explaining strategies, beliefs, procedures, or mere facts (Mayer 2009, pp. 30–51). One reason for choosing "enzyme specificity" in particular is the high interactivity among elements of this concept which makes it a typical subject matter for CLT studies as well (Sweller and Chandler 1994). A third reason is that "enzyme specificity" is a common concept in biology and biochemistry, which are the subject matter of the MER framework.

Perhaps, the first question to answer when developing an assessment instrument for the three perspectives, CLT, CTML, and MER is whether it assesses knowledge and/or understanding of a presented material that is inherently complex. Again, CLT researchers define complex material as one that consists of six or more interacting elements (Sweller and Chandler 1994). The concept of enzyme specificity comprises eight of these – assumed for a novice learner in the field. It states that: (1) Binding of a substrate with (2) a specific chemical structure (3) drives the enzyme to (4) undergo the (5) proper conformational change necessary to (6) align the (7) various catalytic amino acid residues (8) along with this substrate. These several steps along with other underlying processes such as *binding affinity of the substrate to the enzyme* and *enzyme reactivity toward the substrate* collectively define this concept as fairly complex by CTML researchers as well. Recall that CTML considers a lesson as "so complex" if it is "consisting of many steps and underlying processes" (Mayer 2009, p. 80). According to the set educational objectives in this instrument, the learner also has to move back and forth along the hierarchical levels of organization in this biochemical phenomenon between the symbolic level (forms of chemical interaction) and the submicro level (functional groups present in the active site of the enzyme). Hence, complexity in MER terms is also obtained.

Content and Face Validity

Content Validity

We first asked a professor emeritus in biochemistry to examine the clarity of each Item and whether test Items are aligned with the objectives listed in Box 1. He also examined whether the presented data in the test resemble those that might be obtained in real experiments.

Box 1 Concept Objectives

Retention Section (Test Items 1.a-d, 2):
To test if the students have *remembered* the following concepts from class discussion:

R1. Binding of the substrate to its enzyme induces the enzyme to undergo a *conformational change* so as to fit the substrate thus catalyzing the chemical reaction

R2. A substrate that is analogous to the natural substrate of an enzyme is *likely* to:

    R2a. Have a *lower* binding affinity to the enzyme

    R2b. Induce the enzyme to undergo a *different* conformational change

    R2c. Cause the enzyme to function at a *lower* reactivity compared to the natural substrate

    R2d. Have an *improper* orientation with the catalytic amino acids and yield *no* products

Transfer Section (Test Items 3–21):
    To test if the students are able to *transfer* (use) the following concepts:

T1. Main factors affecting enzyme specificity are:

    T1a. The *binding affinity* of the enzyme for a substrate

    T1b. The *reactivity* of the enzyme toward a substrate

T2. Undergoing a *proper* conformational change is necessary for the enzyme to appropriately *align* the substrate with the catalytic amino acids

T3. Catalytic amino acids are responsible for *stabilization of the transition state* of the substrate (i.e. increasing its binding affinity) and/or *increasing the reactivity* of the substrate

Table 1 Subject matter expert rating of item importance

Test item		Met objective	Subject matter expert rating			% E/U
			E	U	Not necessary	
Retention section	1.a	R2.a	6	1	0	100
	1.b	R2.b	5	2	0	100
	1.c	R2.c	4	3	0	100
	1.d	R2.d	4	2	1	85.7
	2	R1	6	1	0	100
Transfer section	1*	Illustrative	3	2	2	–
	2*	Illustrative	3	2	2	–
	3	T1.a	5	2	0	100
	4	T1.a	4	3	0	100
	5	T1.b, T2	5	2	0	100
	6	T1.b, T2	4	3	0	100
	7	T1.a, T1.b, T2	4	3	0	100
	8	T1.b	6	1	0	100
	9	T1.a	6	1	0	100
	10	T1.a	6	1	0	100
	11	T1.b	6	1	0	100
	12	T1.a	6	0	1	85.7
	13	T1.a, T1.b	5	2	0	100
	14	T1.a, T1.b	5	2	0	100
	15	T1.a, T1.b	5	2	0	100
	16	T1.a, T1.b	5	2	0	100
	17	T3	6	1	0	100
	18	T3	5	2	0	100
	19	T3	6	1	0	100
	20	T3	6	1	0	100
	21	T3	5	2	0	100

Note: E = Essential, U = Useful but not essential, *Item is illustrative and hence not rated

After making the necessary edits recommended by this subject matter expert, we asked seven faculty members –five biochemists and two chemists– from different universities to rate each Item as *essential, useful but not essential,* or *not necessary* (Table 1). All test Items were included in the version distributed to the sample of students because 85.7–100% of the responses deemed each Item as *essential/useful but not essential* E/U to assess knowledge and/or understanding of enzyme specificity. We consider these responses as quite supportive to content validity because over 95% of the Items received 100% E/U rating.

Face Validity

To obtain face validity for the cognitive process dimension, four science educators and an educational psychologist examined each test Item as it pertains to one of the six cognitive orders in revised Bloom's taxonomy. All of the five examiners arrived

Table 2 Item specification grid

Test	Cognitive order					
Section	Remember	Understand	Apply	Analyze	Evaluate	Create
Retention	1.a-d, 2	—	—	—	—	—
Transfer*	—	3, 4, 5, 6	17, 18, 19, 20, 21	8, 9, 10, 11, 12	13, 14, 15, 16	7

*Items 1 and 2 in transfer test are only illustrative and hence were not rated

to an agreement on every test Item except a single Item which was eliminated from the final version of the instrument. Table 2 shows the cognitive order of each Item.

Discriminant Validity

Discriminant validity checks whether an instrument can discriminate between groups that it theoretically should discriminate between. Since representational competence is context specific, we theorized that a valid instrument assessing knowledge/understanding of a scientific concept at the tertiary level education (enzyme specificity) should discriminate between science and non-science students. Accordingly, we instructed and tested a sample of 111 college students comprising of 48 science students —biology/biochemistry— and 63 non-science students (medical technology, nursing, or nutrition).

Overall Test

Controlling for the significant difference in pre-test scores ($\alpha = 0.509$), t(109) = 4.69, p < 0.001, One-Way ANCOVA demonstrated that participant's major (science versus non-science) had a significant effect on their overall test score ($\alpha = 0.740$) with science students scoring higher than their non-science counterparts, F(2, 108) = 11.45, p = 0.001. The observed power of this test was 0.918 which indicates that a Type I error is unlikely. Therefore, this assessment instrument satisfies the discriminant validity check because it can distinguish between groups that it theoretically should distinguish between.

Retention Section

Here, discriminant validity of the test section assessing content knowledge *"Remember"* of science and non-science students was examined. As theorized, science students scored higher than their non-science counterparts before instruction ($\alpha = 0.509$), $F(1, 109) = 21.99$, $p < 0.001$. Similarly, they did so after instruction

($\alpha = 0.604$), $F(1, 108) = 14.16$, $p < 0.001$.Therefore, this section satisfies the discriminant validity check. However, poor reliability of pre-test scores is acknowledged ($\alpha = 0.509$).

Transfer Section In this section, discriminant validity of the entire transfer section as well as each of its cognitive orders "Understand through Create" were examined. Given that transferring knowledge is inherently related to the amount of acquired knowledge in the first place, participants' scores on the retention section were used as a covariate while testing transfer performance because of the obtained significant difference in retention. The choice for this measure is statistically supported by the fact that scores on retention significantly predicted transfer scores, $\beta = 1.23$, t(108) = 6.23, p < 0.001. Retained knowledge after instruction explained a significant portion of the variance in transfer performance, $R^2 = 0.265$, F(1, 108) = 38.84, p < 0.001.

Science students demonstrated significantly higher transfer performance than non-science students with a participant's major explaining almost 32% of the variance in transfer performance ($\alpha = 0.706$), $R^2 = 0.319$, $F(2, 107) = 24.69$, $p < 0.001$. Therefore, this section also satisfies the discriminant validity check because it can discriminate between both groups. Again, the transfer section involved '*Understand* through *Create*' Items that are discussed in details below. *F*-tests were conducted for the cognitive orders "*Apply, Analyze,* and *Evaluate*" to test the difference between the two groups. Wilcoxon rank-sum tests were conducted for cognitive orders "*Understand* and *Create*" because *F*-test assumptions of normal distribution were not met ($z = -3.88$ and $z = 5.66$ respectively; $p < 0.05$) and sample sizes were not equal.

Understand In this set of questions, science students demonstrated higher understanding of the presented concept than non-science students ($W = 3118, p = 0.008$).

Apply Here, science students were better able to apply scientific information to a real experimental problem than non-science students $F(2, 107) = 17.27, p < 0.001$.

Analyze In this set of Items, a participant's major had a significant effect on their ability to analyze the results of scientific experiments presented in the test with science students scoring higher than their non-science counterparts $F(2, 107) = 6.25$, $p = 0.003$.

Evaluate Here, participants of science majors showed better ability to critique the given hypotheses than their non-science counterparts $F(2, 107) = 3.09, p < 0.05$.

Create In this question, science students were better able to conceive a novel solution to the given scientific problem than non-science students ($W = 3159, p = 0.008$).

With this set of statistical results, we demonstrated the discriminant validity of each cognitive order/set of Items in the transfer section from *Understand* to *Create* Items.

Item Analysis

Item difficulty index p represents the proportion of test takers who answered the Item correctly. Mathematically, p can range from 0 (none of the test takers answered the Item correctly) to 1 (all test takers answered the Item correctly). Generally, difficulty values below 0.2 are considered very difficult Items, and values above 0.9 are considered very easy Items (Chang et al. 2011). Such Items do not provide valuable information about students' abilities. As displayed in Table 3, difficulty values for the entire test Items fell within this range and therefore were neither too easy nor too difficult for our sampled students.

Discrimination index D demonstrates how well the Item serves to distinguish between test takers based on either an internal or external criterion. For reliability measures, D is computed based on an internal criterion such as total test scores (Aiken 2003, p. 68). This is shown on the rightmost column of Table 3 and shows that all Items are discriminating positively except for Item 16 which was dropped from the test. For validity measures however, D is computed based on an external criterion which, in our case, is the major of the participant – science *versus* non-science students (Aiken 2003, pp. 68–69). Mathematically, D can range from -1 (e.g. all non-science students answered the Item correctly but none of science students did) to $+1$ (e.g. all science students answered the Item correctly but none of non-science students did). In general, D value of 0.2 and above is acceptable. For standardized tests however, D value of 0.3 and above is desirable (Doran 1980). A universal framework for analyzing D does not seem to exist though. For instance, Brown and Abeywickrama (2004) state that, "no absolute rule governs the establishment of acceptable and non-acceptable [D] indices." Yet, with difficulty and discrimination indices being inherently related, D values might be interpreted along with corresponding p values for each Item. Brennan (1972) provides the following criteria for this analysis: (a) Items that discriminate negatively are clearly unacceptable because the lower group outperformed the upper group, (b) Items that discriminate positively are acceptable if the criterion is to differentiate between the two groups, (c) a non-discriminating Item with low p value is not ideal because it is too difficult for both groups, and (d) a non-discriminating Item with high p value is acceptable because both groups are passing the Item.

Accordingly, Items 1.b and 18 are unacceptable because they were negatively discriminating Items ($D < 0.0$). Items 1.c, 1.d, 5, 11, 17, 19, and 20 are acceptable because they were positively discriminating Items ($D > 0.2$). Items 9, 12, and 21 are not ideal because they were non-discriminating with low p values ($p < 0.5, D < 0.2$). The rest of Items are acceptable because they were non-discriminating with high p values ($p > 0.5, D < 0.2$). Based on this Item analysis, all of the test Items were retained except for Item 1.b, 16, and Item 18, which are highlighted by an asterisk in Table 3.

Table 3 Item difficulty and item discrimination indices

Item	p	p for sciences[a]	p for non-sciences[b]	D^c	$D^{\,d}$
Retention section					
1.a	0.55	0.65	0.47	0.18	0.59
1.b*	0.76	0.71	0.81	−0.10	0.24
1.c	0.62	0.77	0.50	0.27	0.55
1.d	0.43	0.58	0.31	0.28	0.66
2	0.61	0.71	0.53	0.18	0.41
Transfer section					
3	0.81	0.88	0.76	0.11	0.34
4	0.81	0.85	0.78	0.08	0.48
5	0.68	0.81	0.57	0.24	0.62
6	0.74	0.79	0.70	0.09	0.55
7	0.22	0.88	0.71	0.16	0.34
8	0.90	0.31	0.14	0.17	0.47
9	0.78	0.94	0.87	0.06	0.21
10	0.61	0.65	0.59	0.06	0.34
11	0.64	0.77	0.54	0.23	0.41
12	0.35	0.40	0.32	0.08	0.24
13	0.62	0.67	0.59	0.08	0.55
14	0.54	0.56	0.52	0.04	0.41
15	0.68	0.75	0.63	0.12	0.59
16*	0.29	0.33	0.25	0.08	−0.07
17	0.38	0.58	0.22	0.36	0.55
18*	0.28	0.25	0.30	−0.05	0.28
19	0.42	0.56	0.32	0.25	0.69
20	0.49	0.73	0.30	0.43	0.69
21	0.41	0.44	0.38	0.06	0.45

[a]Difficulty index computed just for sample of science students
[b]Difficulty index computed just for sample of non-science students
[c]Discrimination index based on the external criterion (participant's major)
[d]Discrimination index based on the internal criterion (overall test score)
*Unacceptable Item

Item 7

Validity and reliability of Item 7, the only free response question, are discussed in this section. Here, we asked students to design an artificial substrate with higher affinity to the enzyme but still receive no enzyme reactivity. We required them to explain their solution plan both in words and in drawings in order to maintain cross-data validity checks (Patton 2002). One students' response is presented (Fig. 1) below to demonstrate how their drawings were used to validate interpreting their verbal responses. Participant_465305: "You could add another bonding site that interrupts the site of the bond to be broken. Interrupts as in stops it from fully closing on it."

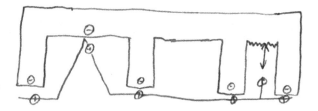

Fig. 1 Drawing for Participant_465305

By referring their verbal response to the drawing, one can tell that this participant conceived the solution through two structural modifications to the natural substrate. To increase binding affinity, they added a negatively charged group to the rightmost side of the substrate "add another bonding site" to utilize the free positive charge on the enzyme. Alongside, they moved the bond to be broken away from the catalytic group to "stop[s] [the bond] from fully closing on [the catalytic group]." This is evident from the up-down open headed arrow in the drawing. This modification was deemed acceptable as it is expected to reduce enzyme reactivity toward the substrate.

Box 2 Rubric for Item 7

Increasing binding affinity:
Any modifications in the structure of the substrate that would increase the number of non-covalent bonds between the enzyme and the substrate are acceptable. This includes but is not limited to:

- Introduction of an additional attractive group on the substrate that would interact with the free rightmost group on the enzyme
- Introduction of additional attractive group(s) that would result in formation of more bonds between a given group on the enzyme and the introduced one(s) along with the already existing bonds that this given group is forming.

Reducing reactivity:
Any modifications in the structure of the substrate that would block/weaken the interaction of the catalytic group with the bond to be broken are acceptable. This includes but is not limited to:

- Displacement of the bond to be broken in a way that prevents access of the catalytic group by any alternative forms of conformational change.

Replacement of the bond to be broken by another that is nonreactive to the catalytic group.

To help reduce potential bias, Item 7 was graded by two independent raters based on a predefined rubric (see Box 2) and examination of absolute agreement resulted in a high intra-class correlation coefficient (0.983).

Gauging Differences in Representational Competence among Students of Different Prior Knowledge

The analysis above showed that the instrument in hand satisfied content and face validities. More importantly, it satisfied discriminant validity with science students distinguished from non-science students by scoring higher in both retention and transfer sections. Although this analysis hints to the role of prior knowledge, it does not provide specific guidance for how to redesign instruction to bridge the gap between the two groups. Particularly, it does not highlight misconceptions that non-science students need to overcome to better interpret/use the provided external representations in the test. With distractor analysis however, this goal is attained along with highlighting misconceptions held by both groups.

Distractor Analysis

Except for Item 12 and Item 16, the majority of science students (mean ± sd, 72% ±12) chose the correct answer of every Item where less non-science students did (mean ± sd, 55% ±180). Item 12 and Item 16 were correctly answered by less than half of test takers, 35% and 27% respectively. However, distractor analysis of these Items indicated that they might be useful for detecting lack of representational competence in both groups. For instance, both Items revealed that 48–56% of test takers acknowledged the contribution of an amino acid to binding affinity although the corresponding representation (k_m) was not even provided in these two Items (distractor B of Item 12 & Item 16).

Other distractors explained differences in representational competence between science and non-science students. For example, distractor A of Item 12 indicated that 21% of non-science students did not understand the term (k_{cat}) that was defined in the problem; compared to only 2% of science students, Table 4. This is evident from the improper usage of k_{cat} to interpret changes in binding affinity when it is used to represent enzyme reactivity.

Yet in another example, 30% of non-science students (*versus* 15% of science students) thought that enzyme reactivity would increase simply because the representation shows a conformational change taking place, with no regard to the improper orientation of the catalytic group (distractor B of Item 5), Fig. 2. This finding again reflects the influence of prior knowledge on understanding representa-

Table 4 Counts (Percentage) of students in each major that chose each answer

Item	Major	A	B	C	D
Retention Section					
1.a	Science	4 (8%)	11 (23%)	32 (67%)*	1 (2%)
	Non-Science	15 (24%)	19 (30%)	27 (43%)*	2 (3%)
1.b	Science	9 (19%)	33 (69%)*	6 (13%)	0 (0%)
	Non-Science	9 (14%)	49 (78%)*	5 (8%)	0(0%)
1.c	Science	3 (6%)	9 (19%)	34 (71%)*	2 (4%)
	Non-Science	12 (19%)	20 (32%)	31 (49%)*	0 (0%)
1.d	Science	4 (8%)	11 (23%)	11 (6%)	30 (63%)*
	Non-Science	8 (13%)	33 (52%)	2 (3%)	20 (32%)*
2	Science	9 (19%)	29 (60%)*	10 (21%)	0 (0%)
	Non-Science	14 (22%)	32 (51%)*	9 (14%)	8 (13%)
Transfer Section					
3	Science	0 (0%)	3 (6%)	43 (90%)*	2 (4%)
	Non-Science	7 (11%)	7 (11%)	48 (76%)*	1 (2%)
4	Science	2 (4%)	2 (4%)	42 (88%)*	2(4%)
	Non-Science	5 (8%)	9 (14%)	47 (75%)*	2 (3%)
5	Science	2 (4%)	7 (15%)	39 (81%)*	0(0%)
	Non-Science	6 (10%)	19 (30%)	35 (56%)*	3 (55)
6	Science	2 (4%)	6 (13%)	39 (81%)*	1 (2%)
	Non-Science	8 (13%)	10 (16%)	42 (67%)*	3 (5%)
8	Science	45 (94%)*	2 (4%)	1 (2%)	0 (0%)
	Non-Science	55 (87%)*	3 (5%)	5 (8%)	0(0%)
9	Science	2 (4%)	44 (92%)*	2 (4%)	0 (0%)
	Non-Science	10 (16%)	44 (70%)*	9 (14%)	0 (0%)
10	Science	5 (10 %)	34 (71%)*	9 (19%)	0 (0%)
	Non-Science	8 (13%)	36 (57%)*	15 (24%)	4 (6%)
11	Science	36 (75%)*	5 (10%)	6 (13%)	1 (2%)
	Non-Science	36 (57%)*	9 (14%)	16 (25%)	2 (3%)
12	Science	1 (2%)	27 (56%)	19 (40%)*	1 (2%)
	Non-Science	13 (21%)	30 (48%)	20 (32%)*	0 (0%)
13	Science	33 (69%)*	7 (15%)	6 (13%)	2 (4%)
	Non-Science	35 (56%)*	21 (33%)	7 (11%)	0 (0%)
14	Science	41 (65%)*	11 (23%)	5 (10%)	1 (2%)
	Non-Science	30 (48%)*	22 (35%)	11 (17%)	0 (0%)
15	Science	4 (8%)	35 (73%)*	6 (13%)	3 (6%)
	Non-Science	12 (19%)	43 (68%)*	8 (13%)	0 (0%)
16	Science	5 (10%)	26 (54%)	15 (31%)*	2 (4%)
	Non-Science	13 (21%)	30 (48%)	15 (24%)*	5 (8%)
17	Science	30 (63%)*	8 (17%)	5 (10%)	5 (10%)
	Non-Science	15 (24%)*	23 (37%)	15 (24%)	10 (16%)
18	Science	6 (13%)	5 (10%)	13 (27%)*	24 (50%)
	Non-Science	7 (11%)	13 (21%)	18 (29%)*	25 (40%)

(continued)

Table 4 (continued)

Item	Major	A	B	C	D
19	Science	3 (6%)	29 (60%)*	9 (19%)	7 (15%)
	Non-Science	17 (27%)	17 (27%)*	14 (22%)	15 (24%)
20	Science	3 (6%)	7 (15%)	2 (4%)	36 (75%)*
	Non-Science	12 (19%)	18 (29%)	14 (22%)	19 (30%)*
21	Science	4 (8%)	13 (27%)	23 (48%)*	8 (17%)
	Non-Science	11 (17%)	17 (27%)	23 (37%)*	12 (19%)

*Denotes correct answer

Fig. 2 *Left* side, shows Transition State-A (TS-A) with the enzyme conforming properly in which the catalytic group © is facing the bond-to-be-broken (*broken line*). *Right* side, shows TS-B with the enzyme conforming in a way that kept the catalytic group away from the bond-to-be-broken

tions as less science students missed the causal relationship between the catalytic group and reactivity of the enzyme.

Gauging Changes in Representational Competence along Problems of Different Cognitive Orders

Several researchers pointed that students' representational competence may change depending on the difficulty of the task (Barnea and Yehudit 2000; Halverson and Friedrichsen 2013; Kozma and Russell 2005). Therefore, it might be of interest to gauge this change through assessment so that instruction with external representations can be tailored to meet a certain set of educational objectives. This is very likely in college education where students are expected to perform high cognitive tasks (e.g. applying concepts and analyzing experimental results) rather than recalling information (Zhao et al. 2014). Below are two examples.

In our instrument, we had "*Understand*" and "*Create*" Items sharing the same representation, which is a submicroscopic representation of an enzyme interacting with the transition state of a given substrate, Fig. 3. In "*Understand*" Items, students had to explain how changes in the structure of the enzyme would change enzyme reactivity and binding affinity of the substrate. In the "*Create*" Item, they had to

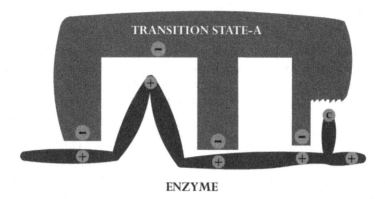

ENZYME

Fig. 3 On *Top*: Transition State-A. *Bottom*: The enzyme. Key: © = catalytic group. *Broken line* = bond-to-be-broken. Circled plus = positively charged site. Circled minus = negatively charged site

perform changes in the structure of the enzyme to obtain certain changes in enzyme reactivity and binding affinity.

Our instrument was able to detect changes in representational competence as students moved along the two problems with their scores dropping significantly from "*Understand*" (0.759 ± 0.286) to "*Create*" Items (0.216 ± 0.341), $t(110) = 15.76$, $p < 0.001$. Obviously, the difficulty here is influenced by the change in cognitive order between the two problems as both problems share the same representation and variables (enzyme reactivity and binding affinity).

In this same instrument, we also had "*Analyze*" and "*Evaluate*" Items sharing the same representations. These are symbolic representations of enzyme reactivity (k_{cat}) and binding affinity (k_m). In "*Analyze*" Items, we asked students to choose the most relevant result to determine the contribution of an amino acid to specificity. In "*Evaluate*" Items, we tested their ability to evaluate a set of claims based on the same results given in "*Analyze*" Items. Again, we detected changes in representational competence as students moved along the two problems with their scores dropping significantly from "*Analyze*" (0.734 ± 0.259) to "*Evaluate*" Items (0.615 ± 0.339), $t(110) = 3.280$, $p = 0.001$. Distractor analysis further supported the claim that this drop in scores stems from representational competence as 11–13% of students mistakenly used k_{cat} as a representation for binding affinity (distractor A of Item 9, Item 10, & Item 12) while 13% mistook k_m as a representation for enzyme reactivity (distractor B of Item 11).

Discussion

In this chapter, we demonstrated how assessment can be employed to detect differences in representational competence among learners of different prior knowledge. We also showed how changes in representational competence can be gauged along problems of various difficulties. Yet, mere detection of differences/changes in representational competence does not provide specific guidance for treatment. Traditional assessment practices, such as Item analysis, sometimes help specify the treatment by specifying the factors influencing competence. This is true for cases where prior knowledge is the main cause of variation in competence as we saw earlier in this chapter. In some other cases, diagnosis need to be performed under theoretical lenses such as CLT, CTML, and the MER framework in order to identify the relevant treatment. Take the *"Create"* Item for instance. Relying on traditional measures, we identified the drop in competence from *"Understand"* to *"Create"* Items, but this was not enough to recommend a treatment that would move more students to engage in such a highly cognitive task. However, upon examining *"Understand"* and *"Create"* Items with a CLT lens, we recognize more interacting elements in the *"Create"* Item than in *"Understand"* Items. For an average learner to answer one of *"Understand"* Items, they need to simultaneously consider the following eight elements:

1. The enzyme is charged, positively; the transition state TS is charged, negatively. The enzyme and the TS mutually attract because they bare opposite charges.
2. The degree of interaction between the TS and the enzyme depends on the number of charged sites on each.
3. The degree of this interaction reflects binding affinity of the TS to the enzyme.
4. © refers to the catalytic group of the enzyme.
5. The broken line refers to the bond-to-be-broken in the TS.
6. For the enzyme to act on the TS, © needs to interact with the bond-to-be-broken.
7. The interaction between © and the bond-to-be-broken depends on the conformation of the enzyme, which is dictated by the interaction of the charged sites.
8. The interaction between © and the bond-to-be-broken reflects enzyme reactivity.

However, to answer the *"Create"* Item, the learner needs to consider these eight elements along with the additional elements below:

1. To increase binding affinity, modify the structure of the TS in a way that would increase the degree of mutual interaction with the enzyme.
2. To reduce enzyme reactivity, modify the structure of the TS in a way that blocks/weakens the interaction between © and the bond-to-be-broken.
3. The measure taken to increase binding affinity should not result in a conformation that enhances enzyme reactivity.

4. The measure taken to reduce enzyme reactivity should not decrease binding affinity.
5. Modification(s) that result(s) in increased binding affinity along with decreased enzyme reactivity comprises a valid answer.

This difference in element interactivity between *"Understand"* and *"Create"* Items is sufficient to suggest a treatment from CLT's repertoire so that the *"Create"* Item is solved with minimal demands on the learner's limited working memory. CLT part-whole sequencing technique, for example, can be employed in which this set of elements is divided into two parts (Van Merriënboer et al. 2006). One part comprises the first three elements that relate to binding affinity. The other part consists of the following four elements that relate to enzyme reactivity. Upon practice, elements comprising each part become automated as single schemas thus freeing space in the working memory necessary to engage in creating a novel solution for the given problem.

It is worth noting however that, this number of interacting elements is certainly an estimate as students in our sample came from different trainings. That is, the number may be less for senior science students as they may be already aware that opposite charges attract and that the degree of this attraction influences affinity, etc. On the other hand, some of these elements may need to be divided into further elements for freshman non-science students, as they may not be aware that the conformation of the enzyme is influenced by the distribution and number of charged sites, etc. In all cases, the "Create" Item remains more demanding than *"Understand"* Items with its additional interacting elements.

Another example comes from *"Apply"* Items. Students' scores on *"Apply"* Items were significantly low (0.423 ± 0.332) compared to *"Analyze"* Items (0.734 ± 0.259, $t(110) = -8.645$, $p < 0.001$) and *"Evaluate"* Items (0.615 ± 0.339, $t(110) = -4.979$, $p < 0.001$) although the latter's belong to higher cognitive orders. Nevertheless, if we examine these Items with the MER lens, we notice that in *"Apply"* Items students had to study charge and polarity (terms) of the presented side chains of amino acids to determine changes in enzyme reactivity and binding affinity (concepts). In other words, they had to vertically translate from terms to concepts. They also had to horizontally translate back and forth among words, presented side chains, and a chemical equation. With *"Analyze"* and *"Evaluate"* Items however, students had no vertical translations as they worked only at the concepts level. Horizontally, they only had to translate between words and symbols. Therefore, the range of needed translations is wider in *"Apply"* Items than in the other two. The question that follows is, which of the two translations, vertical or horizontal, is turning *"Apply"* Items that difficult? With distractor analysis, we found that a higher percentage of non-science students could not perform the needed vertical translation from terms to concepts because they could not properly use the presented information about each amino acid. This is obvious from distractor A of Items 19, 20, and 21; distractor B of Items 17, 18, and 20; and distractor C of Item 20 (Table 4). With this being

known, it would be easy then for CLT and CTML researchers to recommend including the relevant information of the presented amino acids (charge and polarity)
within a pre-training episode. CLT researchers would further argue that the learner
needs to cognitively *automate* this information (*say through a drill-and-practice
exercise within the pre-training episode*) since they recur over and over in *"Apply"*
Items (Van Merriënboer and Sweller 2005). This automation would allow the characteristics of amino acids to be processed automatically rather than consciously in
working memory, thus reducing the difficulty posed by these particular Items
(Pollock et al. 2002).

Certainly, CLT, CTML, and the MER framework are not the only lenses revealing the challenges that face students while learning with external representations.
Nor they are the only ones that offer effective treatments. However, we decided to
restrict our discussion to these three in particular for the complementary function
they offer when taken together. Other theories/frameworks can also be quite informative for studying representational competence. One promising theory is the
Cognitive-Affective Theory of Learning with Media (CATLM) which accounts for
motivational factors that may contribute to learning with representations (Moreno
and Mayer 2007). We find this an interesting field to investigate as some advanced
learners, for instance, may not be motivated to attend to details in a visual representation assuming that it is redundant information, which mistakenly would turn them
as representationally incompetent.

Again, we found that learner's prior knowledge, element interactivity, cognitive
orders of given problems, and degree of needed translations as factors influencing
representational competence. By no means however, we consider these a comprehensive list of the factors that influence representational competence. Instead, we
preferred to discuss data-based ones without denying that other factors may also
apply. One potential factor for example is the difference in cognitive skills among
learners of different majors/occupations. Recall in our sample that we had two
groups of learners, the first group consisted of science students. The second consisted of medical technology, nutrition, and nursing students. In the test, both groups
took two sets of test Items among others. In the first set, they had to analyze given
results, and in the second they had to evaluate claims based on these same results.
Like all other Items, scores of non-science students on *"Analyze"* Items were significantly lower than those of science students, $F(2, 107) = 6.25, p = 0.003$. However,
their scores on *"Evaluate"* Items were statistically ($F(2, 107) = 3.09, p = 0.049$) but
not practically different (Cohen's d is as low as 0.21) from those of science students.
Consequently, and unlike the other group ($t(48) = 2.908$, p $= 0.006$), non-science
students demonstrated similar representational competence as they moved from
"Analyze" to *"Evaluate"* Items, $t(62) = 1.900$, *ns*. This observation may be referred
to the training that health sciences students receive, which includes *evaluating* data-
driven claims based on defined criteria. For example, they are used to evaluating
claims such as "diabetes *may contribute* to dehydration" based on relevant facts
such as "diabetic patients experience excessive urination." Therefore, their honed
evaluation skills might have moderated the potential change in representational
competence that could have happened because of the different cognitive orders of

these two problems. Although the data generated by the given instrument does not statistically support this claim ($F(1, 109) = 0.486$, ns), it would be of interest to further study the role of acquired/honed cognitive skills in influencing changes in representational competence.

The described approach to assessment in this chapter has the potential to improve the design of instruction with external representations as it accounts for cases where lack of/differences in representational competence is the main cause of failure in instructional intervention. Logistically, such multiple-choice formatted assessment instruments may be administered and scored automatically by the ubiquitous computers available in classrooms and research labs. Researchers would then directly interpret the electronically generated difficulty index, criterion-based discrimination indices, and distractor analysis along with overall test scores. Arguably, this automatic, objective, and time-saving approach to assessment might be preferred by many researchers over the manual time-consuming grading of free response questions (DeLeeuw and Mayer 2008). Finally, we acknowledge that the results obtained and the corresponding conclusions reached here need to be replicated on a different sample of students and different presented material. Accordingly, we anticipate this work to be a catalyst for further research in this direction and hence consider this chapter as a preface for further literature.

References

Aiken, L. R. (2003). *Psychological testing and assessment*. Boston: Allyn and Bacon.

Anderson, L. W., Krathwohl, D. R., Airiasian, W., Cruikshank, K. A., Mayer, R. E., & Pintrich, P. R. (2001). A taxonomy for learning, teaching and assessing: A revision of Bloom's taxonomy of educational outcomes: Complete edition.

Barnea, N., & Yehudit, J. D. (2000). Computerized molecular modeling-the new technology for enhancing model perception among chemistry educators and learners. *Chemistry Education Research and Practice, 1*(1), 109–120.

Brennan, R. L. (1972). A generalized upper-lower item discrimination index. *Educational and Psychological Measurement, 32*, 289–303.

Brown, H. D., & Abeywickrama, P. (2004). *Language assessment. Principles and Classroom Practices*. White Plains: Pearson Education.

Chang, M., Hwang, W. Y., Chen, M. P., & Mueller, W. (Eds.). (2011). *Edutainment technologies. Educational games and virtual reality/Augmented reality applications: 6th International Conference on E-learning and Games, Edutainment 2011, Taipei, Taiwan, September 7–9, 2011, Proceedings* (Vol. 6872). Springer.

Dacosta, B. (2008). The effect of cognitive aging on multimedia learning (Doctoral dissertation, University of Central Florida Orlando, Florida).

DeLeeuw, K. E., & Mayer, R. E. (2008). A comparison of three measures of cognitive load: Evidence for separable measures of intrinsic, extraneous, and germane load. *Journal of Educational Psychology, 100*, 223–234.

Doran, R. L. (1980). *Basic measurement and evaluation of science instruction*. National Science Teachers Association, 1742 Connecticut Ave., NW, Washington, DC 20009 (Stock No. 471–14764; no price quoted).

Griffard, P. B. (2013). Deconstructing and decoding complex process diagrams in university biology. In *Multiple representations in biological education* (pp. 165–183). Netherlands: Springer.

Halverson, K. L., & Friedrichsen, P. (2013). Learning tree thinking: Developing a new framework of representational competence. In *Multiple representations in biological education* (pp. 185–201). Netherlands: Springer.

de Jong, T. (2010). Cognitive load theory, educational research, and instructional design: Some food for thought. *Instructional Science, 38*(2), 105–134.

Kalyuga, S. (2007). Expertise reversal effect and its implications for learner-tailored instruction. *Educational Psychology Review, 19*(4), 509–539.

Klahr, D., & Robinson, M. (1981). Formal assessment of problem-solving and planning processes in preschool children. *Cognitive Psychology, 13*(1), 113–148.

Kozma, R. B., & Russell, J. (1997). Multimedia and understanding: Expert and novice responses to different representations of chemical phenomena. *Journal of Research in Science Teaching, 34*(9), 949–968.

Kozma, R., & Russell, J. (2005). Students becoming chemists: Developing representational competence. In *Visualization in science education* (pp. 121–145). Dordrecht: Springer Netherlands.

Madrid, R. I., Van Oostendorp, H., & Melguizo, M. C. P. (2009). The effects of the number of links and navigation support on cognitive load and learning with hypertext: The mediating role of reading order. *Computers in Human Behavior, 25*(1), 66–75.

Mautone, P. D., & Mayer, R. E. (2001). Signaling as a cognitive guide in multimedia learning. *Journal of Educational Psychology, 93*(2), 377.

Mayer, R. E. (2009). *Multimedia learning* (2nd ed.). New York: Cambridge University Press.

Mayer, R. E., Mathias, A., & Wetzell, K. (2002). Fostering understanding of multimedia messages through pre-training: Evidence for a two-stage theory of mental model construction. *Journal of Experimental Psychology: Applied, 8*(3), 147.

Moreno, R. (2004). Decreasing cognitive load for novice students: Effects of explanatory versus corrective feedback in discovery-based multimedia. *Instructional Science, 32*(1–2), 99–113.

Moreno, R., & Mayer, R. (2007). Interactive multimodal learning environments. *Educational Psychology Review, 19*(3), 309–326.

Paivio, A. (1986). *Mental representations: A dual coding approach.* Oxford, England: Oxford University Press.

Patton, M. Q. (2002). *Qualitative research and evaluation methods.* John Wiley & Sons, Ltd.

Pollock, E., Chandler, P., & Sweller, J. (2002). Assimilating complex information. *Learning and Instruction, 12*(1), 61–86.

Schönborn, K. J., & Bögeholz, S. (2013). Experts views on translation across multiple external representations. In *Multiple representations in biological education* (pp. 111–128). Dordrecht: Springer Netherlands.

Stull, A. T., & Mayer, R. E. (2007). Learning by doing versus learning by viewing: Three experimental comparisons of learner-generated versus author-provided graphic organizers. *Journal of Educational Psychology, 99*(4), 808.

Sweller, J., & Chandler, P. (1994). Why some material is difficult to learn. *Cognition and Instruction, 12*(3), 185–233.

Treagust, D. F., & Tsui, C. Y. (Eds.). (2013). *Multiple representations in biological education.* Dordrecht: Springer Netherlands.

Treagust, D., & Tsui, C. (2014). General instructional methods and strategies. In N. Lederman & S. Abell (Eds.), *Handbook of research in science education* (1st ed., p. 312). New York: Routledge.

Van Merriënboer, J. J., & Sweller, J. (2005). Cognitive load theory and complex learning: Recent developments and future directions. *Educational Psychology Review, 17*(2), 147–177.

Van Merriënboer, J. J., Kirschner, P. A., & Kester, L. (2003). Taking the load off a learner's mind: Instructional design for complex learning. *Educational Psychologist, 38*(1), 5–13.

Van Merriënboer, J. J., Kester, L., & Paas, F. (2006). Teaching complex rather than simple tasks: Balancing intrinsic and germane load to enhance transfer of learning. *Applied Cognitive Psychology, 20*(3), 343–352.

Zhao, N., Wardeska, J. G., McGuire, S. Y., & Cook, E. (2014). Metacognition: An effective tool to promote success in college science teaching. *Journal of College Science Teaching, 43*(4), 48–54.

Improving Students' Representational Competence through a Course-Based Undergraduate Research Experience

Chandrani Mishra, Kari L. Clase, Carrie Jo Bucklin, and Kristy L. Daniel

Introduction

The major goal of science education is to develop science literacy (National Research Council 1996; Rutherford and Ahlgren 1990). One component of science literacy is the ability to use common representations of scientific concepts and phenomena, such as protein structures and biochemical reactions (Harle and Towns 2013), DNA diagrams (Patrick et al. 2005; Takayama 2005), molecular phenomena (Harle and Towns 2010; Kozma and Russell 2005), and phylogenetic trees (Halverson 2010; Halverson 2011; Baum et al. 2005; Dees et al. 2014; Matuk 2007). Visual representations are critical for communicating abstract science concepts (Patrick et al. 2005; Gilbert 2005b; Mathewson 1999). In science, visual representations are used to display data, organize complex information, and promote a shared understanding of scientific phenomena (Kozma and Russell 2005; Roth et al. 1999). These representations are often used to present multiple relationships and processes that are difficult to describe or observe. Various forms of representations can support an understanding of different, yet overlapping aspects of a phenomenon or entity. Representations play a key role in mathematics, geography, and science (Cuoco and Curcio 2001; Gilbert 2005a), especially in biology, particularly with genetics and evolution. High levels of visualization skills are linked to creativity, not

C. Mishra (✉) · K. L. Clase
Purdue University, West Lafayatte, IN, USA
e-mail: klclase@purdue.edu

C. J. Bucklin
Southern Utah University, Cedar City, UT, USA
e-mail: carriebucklin@suu.edu

K. L. Daniel
Texas State University, San Marcos, TX, USA

© Springer International Publishing AG, part of Springer Nature 2018   177
K. L. Daniel (ed.), *Towards a Framework for Representational Competence in Science Education*, Models and Modeling in Science Education 11,
https://doi.org/10.1007/978-3-319-89945-9_9

only in the arts, but also in science and mathematics (Shepard 1988) and individual case studies support the connections among creativity, scientific discoveries and spatial ability.

Representational Competence

Communicating with representations is a critical aspect of being a scientist (Trumbo 1999; Yore and Hand 2010). It is essential to understand how to use and interpret discipline specific representations to aid in this communication. We refer to this concept as representational competence (Halverson and Friedrichsen 2013). Representational competence is the ability of an individual to understand and use representations when explaining complex phenomenon (Halverson and Friedrichsen 2013; Kozma and Russell 2007). Frameworks for representational competence or fluency have been described in chemistry (Kozma and Russell 2005), mathematics (Meyer 2001), biology (Halverson and Friedrichsen 2013), and biochemistry education (Anderson et al. 2012; Harle and Towns 2012, 2013).

In the original framework (Kozma and Russell 2005), the initial level of representational competence is achieved, "when asked to represent a physical phenomenon, the person generates representations of the phenomenon based only on its physical features. That is, the representation is an isomorphic, iconic depiction of the phenomenon at a point in time" (Kozma and Russell 2005, p. 132). This level of competency focuses on a person's ability to generate representations, not just their ability to make sense of a representation. Halverson et al. (2011) found that student errors in prior knowledge interfered with the process of gaining representational competence. For example, when students were familiar with the organisms on the phylogenetic tree, they used their knowledge of physical and ecological similarities rather than the evolutionary information represented in the structure of the phylogenetic tree. A framework was needed to account for students who fail to use representations to answer questions or fail to generate representations to communicate their knowledge. These two were critical for capturing differences in students' development of representational competence in biology within the context of tree thinking. As a result of this initial study, Halverson and Friedrichsen (2013) proposed a new, tentative, empirically-based framework for representational competence in biology education. We further tested this proposed representational competence framework in our study for its applicability to learn with representations in other contexts of biology, such as genomics.

Student learning with representations is well documented in chemistry, physics, geography and science education in general (e.g., Ferk et al. 2003; Chi et al. 1981; Kozma and Russell 2005; Peterson 1994; Tytler et al. 2013). Although, use of representations to learn in Biology have been noted (Halverson and Friedrichsen 2013; Anderson et al. 2012; Won et al. 2014), there is limited research investigating how students gain representational competence in biology (see Gilbert 2005a; Gilbert and Treagust 2009), specially in the field of genomics. Discovery in modern biology occurs beyond a wet lab bench—authentic biological discovery also relies on discovery in silico and occurs within the virtual environment of a computer. Representations of genomes are abstract and are often displayed in computer software and web browsers that include distinct symbols and pictures for different features of the genome--- i.e. boxes for putative proteins and arrows to depict direction of a biological process (Shaer et al. 2012). Thus, experts must readily work with multiple abstract representations of genomes at different scales that require the use of technology for inquiry and discovery (Shaer et al. 2010). As students acquire knowledge and experiences that move them along the novice-to-expert continuum, they, too will be faced with the challenges that experts face in this field.

The amount of information generated from genome sequencing and the size of genomes of organisms has resulted in a challenge for visualizing the information to understand and make sense of the data and also facilitate comparative analysis among genomes. Standardization of genome representation and genomic data is an issue for scientists, including the best way to visualize the data in hopes of working with the data, making conclusions, observations, and posting hypotheses for future work (Sterk et al. 2006). Scientists grapple with the best way to represent a genome for annotation and standards are updated often as the information becomes more extensive and technology improves (Pruitt et al. 2011).

> One way to conceptualize undergraduate education is as a process of moving students along the path from novice toward expert understanding within a given discipline. To achieve this goal, it is important to begin by identifying what students know, how their ideas align with normative scientific and engineering explanations and practices (i.e., expert knowledge), and how to change those ideas that are not aligned (Singer et al. 2014, p.57).

Developing representational competence promotes thinking and acting like a scientist, students to move away from their novice understanding toward an expert understanding of the subject depicted in the representations. For this research study, we explored how students' thinking changed about genomes by examining students' representations and associated explanations. We approached our study through the perspective that the type of representation in combination with how students talked about genomes was needed to represent a scientific way of thinking about it. Thus, to determine representational competence in accordance with Halverson and Friedrichsen (2013), we looked at the type of representation in combination with students' explanations.

We are using the Representational Competence framework as described by Halverson and Friedrichsen (2013) to guide this study. The authors identified seven levels of representational competence associated with phylogenetic trees. The levels range from no use of representations to expert use of representations. Based on analysis of our pilot data, we adapted these levels to describe students' level of representational competence with annotated genomes. We examined student development of representational competence within the biological context of the genome. As summarized and discussed by Waldrip and Prain (2012), there are two main perspectives of research in representations: examining students' learning with representations and examining student-generated representations. We focused primarily on the use of student-generated representations to reveal their thinking about biological content, processes and scientific literacy.

Methodology

Research Question

We addressed the following research question:

- In what ways do students' representational competence change after participating in a CURE (Course-Based Undergraduate Research Experience)?

Participants

All participants were undergraduate STEM majors enrolled in a course-based undergraduate research experience implemented at a mid-western, public university. Data was collected from the CURE two-course series. The CURE experience was implemented through participation in the Howard Hughes Medical Institute's (HHMI) Science Education Alliance-Phage Hunters Advancing Genomics and Evolutionary Science (SEA-PHAGES) program (Harrison et al. 2011; Jordan et al. 2014). The CURE project is a two semester course sequence that includes one semester of wet lab and one semester of in silico lab. In the wet lab course, students isolate samples from the environment and use microbiology techniques to capture a virus, more specifically a bacteriophage (phage) that infects a specific host strain of bacteria. Then, students spend the rest of the semester using molecular biology techniques to characterize their phage and isolate the DNA of the entire phage genome. The isolated genomes from the viruses are subsequently sequenced between semesters. Students spend the final semester of the CURE experience working with authentic bioinformatics software (i.e. DNAMaster, NCBI BLAST, Phamerator, and GeneMark) to annotate the genome by identifying putative proteins and their functions encoded within the nucleotide sequence. The final genome annotation is submitted online to GenBank at the National Center for Biotechnology Information after final quality control processing by the SEA-PHAGES Mycobacteriophage Annotation Review Team (SMART). Participants for this study included fifty two students from the first semester (Wet Lab) and thirteen students from the second semester (In Silico Bioinformatics) courses.

Data Collection

We collected data from multiple sources including both pre/post course series questionnaires focused on assessing how students make sense of, use, and generate biological representations and individual, semi-structured student interviews. Qualitative data was gathered from student responses on generated representations provided on pre-post questionnaires, and/or interview transcripts. The pre-post questionnaire consisted of three open-ended questions. The first open-ended question asked students to draw how they would represent an annotated genome, the second asked students to describe their representation, and the third asked students to describe the purpose/function of a genome. The interviews were conducted at the end of each semester. Participants were asked about their understanding of annotated genomes and then asked to generate a representation that demonstrated how they visualized an annotated genome. Students were then asked to describe their representation and explain where their ideas about annotated genomes came from. Alternatively, students completed a post questionnaire if they did not participate in the interview.

Data Analysis

We used both quantitative and qualitative analysis methods. We measured students' level of representational competence as fits with the theoretical model described by Halverson and Friedrichsen (2013) using both the student drawing and the explanation provided. Data was collected and analyzed from both the student-generated representation and the discussion, either written on the questionnaire or verbal during the interview, to assign levels of representational competence. For the qualitative portion of our analysis we used a deductive method to code student representations. We dual coded based upon both the student drawn representation and the verbal or written description that revealed the students' use and understanding of representations throughout the course. We grouped similar drawings with descriptions into students' levels of representational competence with annotated genomes using the levels of representational competence as described by Halverson and Friedrichsen (2013). For the quantitative analysis, we determined the overall frequency of students' level of representational competence by combining all of the representations for each level (Fig. 1).

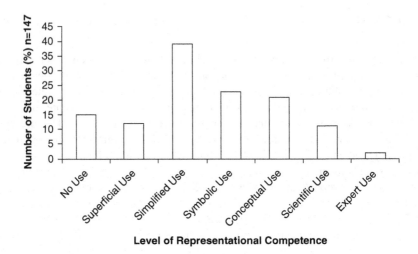

Fig. 1 Graph of frequency distribution of representations

Findings

Students' Level of Representational Competence with Annotated Genomes

We found all seven levels of representational competence with annotated genomes that were previously defined and characterized for phylogenetic trees (Halverson and Friedrichsen 2013). The majority of students' representations were categorized as Level 3: Simplified Use (39%), followed by Level 4: Symbolic Use (23%) and Level 5: Conceptual Use (21%). Very few students (2%) had a representational competence of Level 7: Expert Use (Fig. 1).

Level 1: No Use of Representation

Representations were classified in this category if students did not attempt to draw an annotated genome, did not attempt to talk through what an annotated genome could be, could not describe an annotated genome during their interview, or stated that they did not know what to draw. During his interview, Logan stated, "I honestly have no idea what an annotated genome looks like." As part of her interview Tabitha said, "I don't really know what that is… I don't know what annotated means in the context stuff… I don't really know what a genome is… I thought that genomes contain genetic information." This along with a basic description of chromosomes "it has something to do with genetic information" shows that Tabitha does not have an understanding of what an annotated genome is. Tyler stated "I remember learning about genomes in high school Biology but I do not remember much about them. I think they have to do with DNA and maybe bacteria" and Liz stated, "I don't really

Fig. 2 (**a**) Tyler's representation and (**b**) Liz's representation

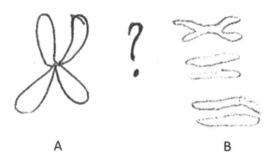

A   B

know how to represent a genome with a picture" (Fig. 2). Similarly Rex said that "I don't really know what an annotated genome is... if I did know what it was I would picture it." Lisa is considered part of this category because she did not attempt to draw anything during her interview and stated "I have no clue."

Level 2: Superficial Use of Representation

This level of representational competence refers to students' creation of representations based on superficial understanding and not connected to in-depth knowledge. Students were classified in this category if they created a literal image to illustrate their understanding of the scientific content and may include appropriate terms but did not provide an accurate explanation with conceptual understanding. Many students in this level also did not have a molecular understanding of the biological concept and instead focused superficially on the macro level of the biological system that they were working with in the laboratory course undergraduate research experience, namely the viruses (bacteriophages) and their host (bacteria). They also appeared to think about an annotated genome as a visible entity that could be physically observed as a whole like the concrete bacteriophage they observed in the laboratory with an electron microscope. Both the student generated drawing supplemented with either verbal or written descriptions, revealed their level of understanding. Jennifer stated "I know phage as having a head and a tail, so that is how I see it" And Mitch stated, "It is a picture of bacteria. The head contains the infected bacteria DNA" (Fig. 3).

Additionally, Anita tried to describe how she thinks about annotated genomes and draw a representation for an annotated genome. She drew an amoeba looking design and wrote "ABC" during her interview (Fig. 4). In her discussion, she focuses more on the regions of DNA, such as promoters, rather than an annotated genome and her representation is literal. She described an annotated genome as "when you have a sequence of bases and stuff, you know this part is an exon and this part is a promoter side. When you have information of what this part of a sequence... that is what I think it is.... This [her drawing] can be an intron." Although she did mention appropriate terms for the biological content area, her explanations are superficial and she struggles to accurately describe them and conceptually connect them to her representation.

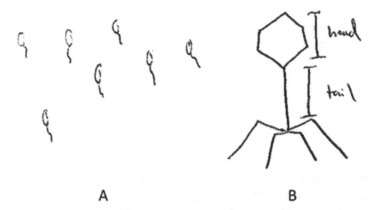

A   B

Fig. 3 (**a**) Jennifer's representation and (**b**) Mitch's representation

Fig. 4 Anita's representation

Level 3: Simplified Use of Representation

This level of representational competence refers to students' creation of representations based on simple but correct understanding of the concept. Students in this category created representations and discussed components of the biological concept with accuracy, but the representations and the explanations were not complete (e.g. drawing chromosomes without annotations; drawing a double helix with base pairs, and drawing karyotypes). Students were also classified in this category if they focused solely on general aspects or properties of genomes, such as the base-pair bonding that allows double helical DNA to hold its shape or drew chromosomes or some other simplified genetic structure to depict an annotated genome. Students at this level do not demonstrate a complete understanding of an annotated genome verbally and do not draw a representation that illustrates all of the components of a genome. Brent drew a single, unduplicated chromosome and described his drawing as "Chromosomes, containing the genetic information necessary for an organism to survive." This student correctly drew an unduplicated chromosome and considers a chromosome to include genes needed for an organism's survival, but not necessarily an organism's entire genetic material (Fig. 5).

Students also focused on their prior knowledge of base pairs and double helix structure in their representations of genomes for this level. Sarah stated "I'm thinking genome, so DNA....there's the backbone, some base pairs, this would be like the TATA box" (Figure 6A). Additionally, Noah and Erica described a genome as a

Fig. 5 Brent's
representation

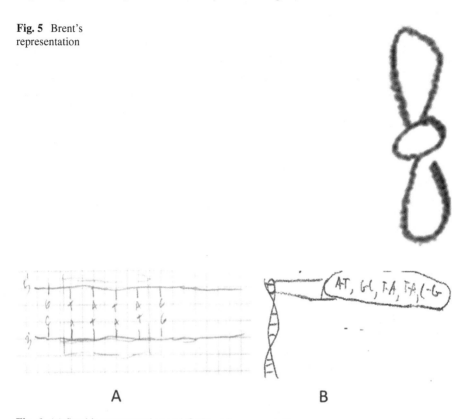

Fig. 6 (**a**) Sarah's representation and (**b**) Noah's representation

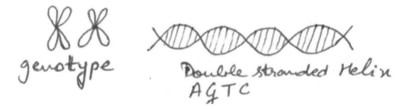

Fig. 7 Billy's representation

base pair sequence, but neglected to describe or draw the annotated portion. Noah described his drawing as simply "a genome is the order in which the four base pairs are organized in the DNA of the organism" (Fig. 6B).

Similarly, Erica stated that the genome is the sequence of bases used to produce proteins.. Billy stated, "It shows a double stranded helix which is held together by bonds formed between the different bases A,G,T,C, and there is also a genotype shown on the left that is the combination of genes from two different sources" (Fig. 7). None of these students (Sara, Noah, Erica, or Billy) accurately described an annotated genome, instead confusing it with a single gene.

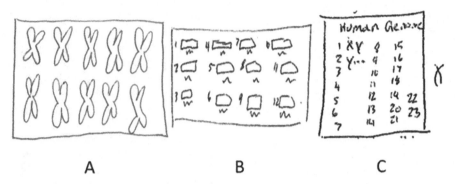

Fig. 8 (**a**) Aaron's representation; (**b**) Sally's representation; and (**c**) Mark's representation

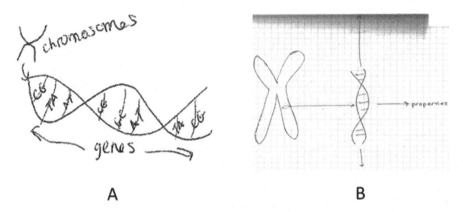

Fig. 9 (**a**) Kathy's representation and (**b**) Dee's representation

Additionally, some students drew representations of karyotypes (Fig. 8) and described them as either chromosomes (in the case of Aaron and Sally) or genotypes (in the case of Mark). Students went on to attempt to connect their drawing to an individual's entire DNA sequence (Aaron) or simply state that their drawing shows what makes up a person (Sally and Mark). For example, Aaron stated, "the genome is the entire DNA sequence. My image shows the chromosomes that contain the DNA sequence." Sally said, "each picture represents a given chromosome and shows what makes it up." and Mark explained, "in my early Biology classes we were told that there were 23 or 26 types genomes in a human. These genomes create who you are."

In some instances students generated representations that seem more advanced that other level 3 representations (i.e. use of multiple scales) but are not advanced enough to be classified completely as level 4 representations (i.e. students use a conceptual foundation). Students may be attempting to make connections between genotypes and genomes, but do not finish making the connection between their drawings and the underlying concepts of annotating genomes. Thus, we classified those representations as level 3 (Fig. 9). Kathy said that, "Genome has chromosomes and genes that contain instructions for making proteins," and her drawing has both a chromosome and a double-helix, but she does not elaborate on how chromosomes and DNA are related to each other or how they work together as instructions for making proteins. Additionally, Dee drew a chromosome with a zoomed in portion that is a double helix. Dee stated that, "I picture a chromosome... this is a chromosome with a bunch of genes, the double helix DNA that's the smallest part. A genome is somewhere in between, the whole strand of DNA is the genome which will express the properties." Dee initially draws a chromosome because that is the first thing she visualizes, next she thinks through the components of a chromosome and then describes DNA strands as being the genome. She later states that she does not know what the annotated part means. Both students use appropriate levels of scale in their representations, but they do not finish making the connection among the different scales and the underlying concepts of genome annotation in either their description or representation.

These simplified representations can be improved by adding annotations or symbols which could also be used to explain the concept and the underlying phenomena. It is discussed more in the next level of representational competence.

Level 4: Symbolic Use of Representation

This level of representational competence refers to students' revealing of their conceptual understanding using representations consisting of symbols or in an annotated form. In this category, students' representations are based on a conceptual understanding of the phenomena, genome annotation, but their ability to link their conceptual understanding to function is incomplete. Students demonstrate that they know that genome annotation is an important concept and it is somehow related to other terms like DNA and nucleotides but they are still not building cohesive connections among the parts. Students know that annotations exist but it is not clear that they know why or what they mean, kind of like DNA is nucleotides. Students know the terms like artifacts, collected facts but not linking them conceptually and building connections among the components----like isolated circles without connections in a system of interrelated working parts that has now been expanded to include genome annotation as a part; however, their understanding is flawed. Students know that annotating includes genes but not connecting deeply to function and sometimes their explanation or representation of genome annotation is inaccurate.

Representations were classified in this category when the drawings referred to the students' conceptual understanding of genetics and genomes, but had superficial descriptions in that they are starting to connect to annotation, but not in a deep conceptual manner or they provided an incorrect description or representation. Two students, Kaleigh and Zane, drew circular DNA and they all described their drawings using DNA or DNA sequence (Fig. 10). Kaleigh drew her circular DNA as a plasmid double helix with a zoom-in box showing how the nucleotides would line up, whereas Zane simply drew a circle with shaded regions that were not labeled. Kaleigh described her image as a DNA sequence. Zane described his drawing as a

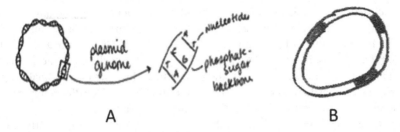

Fig. 10 (a) Kaleigh's representation and (b) Zane's representation

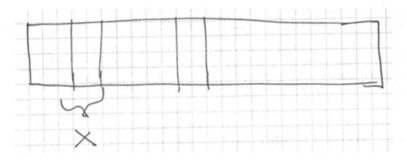

Fig. 11 Emily's representation

"circular, double stranded DNA. The shaded areas are locations of the genome that code for certain genes."

Students' representations may still include linear DNA sequences or chromosomes at this level, but students now start to recognize annotation as assigning function to different portions of the genome. For instance, Emily recognized that the genome was something studied and different portions were marked or annotated as doing specific things or contributing to particular functions. She stated, "I am thinking about genome as all the genes for one organism….there are genes and you can represent them as a box….certain portions of the gene represent something x. I don't know what they represent. So there are just different sections that represent different things" (Fig. 11).

In order to develop a complete conceptual understanding, however, one must be able to relate the system to its function or demonstrate an understanding of the phenomena.

Level 5: Conceptual Use of Representation

This level of representational competence refers to students' creation of representations based on accurate conceptual understanding, however being unable to connect their representations to the conceptual understanding of the phenomena. Students were classified in this category if they demonstrated an understanding of genome

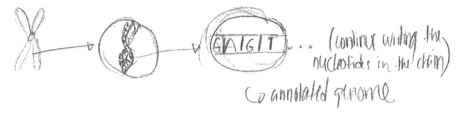

Fig. 12 Jack's pre-representation

annotation but were unable to combine the information in their representations with their accurate conceptual understanding of annotated genomes. Students discuss evaluating evidence for genome annotation or draw representations that include evidence that is evaluated during the process of annotating a genome but they do not connect the evidence to the bigger picture or the rationale for genome annotation----like not being able to see the forest through the trees.

For instance, Samantha stated, "The image shows the genes in the chromosome. The marks are those genes that have been researched and understood." She is attempting to describe that when a scientist annotates a genome, they have researched the genes associated with different traits, however her drawing depicts the 'annotations' on a single chromosome rather than a long portion of DNA. This student conceptually understands the connection between annotated genomes and genes and she was able to express it verbally, even though she was unable to accurately represent that in her drawing. This student is categorized as level 5 rather than level 4 because Samantha is attempting to describe how scientists know where to place the annotations rather than simply stating they exist. Similarly, Jack drew his idea of annotation on a single chromosome during his first interview. However, he incorporated the idea of scale into his drawing to represent where the annotations corresponded to in an individuals' DNA (Fig. 12). Jack stated, "First, I drew a chromosome since it is what contains DNA at the "macro" level (using nucleotides as a point of reference). Then, I kept specifying my drawing until I got to the specific nucleotide sequences that an organism might have."

Students at this level also begin to recognize that scientific representations have underlying data and evidence that is used to create them, even if the data is not always included explicitly in the representation. Students are also beginning to make connections to other components in the system, such as proteins, as they describe their representation (Fig. 13). For instance, Tina stated:

> "The picture is trying to represent gene calls within a genome. There are two forward genes and a reverse gene. If these genes were clicked on, info about the genes and their functions, start codons, SD scores, coding potential, etc. is given about each gene. A genome serves the purpose of functioning, essentially. A genome includes all the genes in your DNA. Genes produce proteins with specific functions. Without a genome, no functioning proteins would be made."

Representations drawn and described by students in this category are often restricted by the representations provided by the bioinformatics programs the stu-

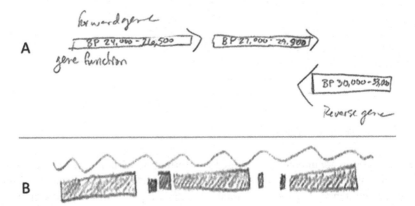

Fig. 13 (**a**) Tina's representation and (**b**) Ella's representation

dents used to annotate the genome during the course. The software representations are evident in the student representations and students' understanding is locked into the technology provided in class for their representation, as revealed in both the representations they create and the descriptions they provide. Some students even discuss the software and "clicking on (per quote from Tina above)" a certain aspect of their representation to reveal additional information or views. The knowledge may be considered brittle----the students seem constrained by the bioinformatics programs used to annotate the genome and they are not integrating these representations from technology into their mental map of representations and deeper conceptual understanding of other relevant biological concepts within the system. Additionally, Ella stated:

> "The boxes represent the gene calls corresponding to the genetic sequence. The calls are based on numerous factors, like the coding potential. The purpose of a genome is to contain all of the information necessary to survival. By better understanding them, we can better understand how organisms work, what genes code for what proteins, the genetic basis for various conditions, and more."

All of the aspects of the representations in this category (i.e. changing scale, connecting system components, recognizing the use of data and uncertainty) demonstrate a way of thinking that is more advanced than a novice learner. The individuals who created these drawings, however, focused narrowly on the application of how to create gene calls and the information found in the computer programs or bioinformatics software rather than how genome annotation relates to their conceptual understanding of genes and genomes.

However, to exhibit a level of representational competency similar to a scientist, one must be able to use their representations to clearly explain the concept or visualize the big picture.

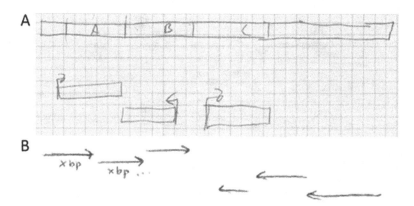

Fig. 14 (**a**) Randy's representation and (**b**) Leah's representation

Level 6: Scientific Use of Representation

This level of representational competence refers to students' creation of representations based on accurate conceptual understanding like the one before, but also can connect their representations to the conceptual understanding of the phenomena. Representations are classified in this category if students are able to draw a *single*, accurate representation and are able to relate their drawing to their accurate conceptual understanding of genes and genomes. Students in this category may also draw a representation that mimics software programs (Fig. 14) but in contrast to Level 5, they now connect their representation accurately to a conceptual understanding of a genome. For instance, Randy's drawing is a series of connected boxes. The boxes are part of DNA and are each labeled as A, B, and C. The labeled boxes are "genes and this is a genome... an annotated genome, you figure out all the genes that are within that sequence genome that are known to do something that somebody figured it out." Leah drew two sets of three arrows pointing at each other. In her description she stated, "I think about the Apollo or GBrowse programs we used... these are good ways to visualize where genes are located in the genome." Leah is drawing on her recollection of her work with the computer to shape how she visualizes an annotated genome. Randy relates his drawing to his conceptual understanding of genomes and Leah relates her drawing to her experience with the computer program. Both students have an accurate conceptual understanding of genomes and a practical understanding of annotating genomes. Students demonstrate an advanced level of technical expertise in the process of assigning and calling genes but they are not making a deeper connection to function and how the technology is enabling them to make these connections. Different levels exist in the biological system and technology is integrated throughout. There is an understanding that students must build among simple nucleotide sequence, a putative gene, the proposed function of that annotated gene and the use of technology along all aspects of the spectrum. Level 6 includes a link of genome annotation to sequence and function but the students did not demonstrate an understanding for the rationale of why the link to

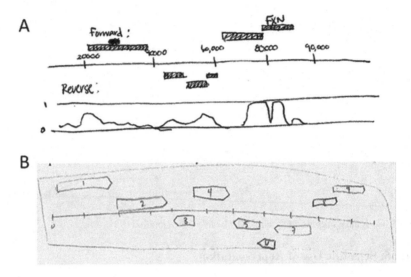

Fig. 15 (**a**) Lilly's representation and (**b**) Lexie's representation

annotation and function was an important step. Finally linking the process of annotation to an understanding of function, with the use of multiple representations, was Level 7.

Alternatively, some students' drawings are classified as level 6, because they only use one representation (what she would see in Apollo or GBrowse) to describe their understanding of annotated genomes (Fig. 15). For instance, Lilly described her representation as, "demonstrating gene calls that have been made, both forward and reverse. This photo also demonstrates that genes can overlap if both are on the forward/reverse side. I also demonstrated coding potential and function." Lexie stated:

> "This image is similar to an Apollo output of a genome. It contains both forward and reverse genes. Each section represents a called gene. Above the grid are forward genes and below the grid are reverse genes. The purpose of a genome is to code for proteins/traits of an organism. It contains DNA which provides instruction for the construction of an organism."

Both Lilly's and Lexie's drawings represent a change in thinking from one of novice toward expert. Both are using data gleaned from Apollo and GBrowse to support their understanding of annotated genomes in the same manner that scientists use data to support their claims. Some students demonstrated a conceptual understanding of the connection between annotated genes and function but they only provided a single representation that mimicked technology. They did not elaborate with more than one representation and the representation was limited to the classroom technology representation. In order to have an expert understanding of a genome, however, scientists and students must be able to make connections to multiple components in the system, use evidence or data to support their representation, and use multiple representations to describe their understanding.

Fig. 16 Bruce's
representation

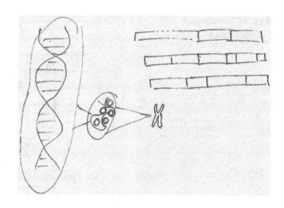

Level 7: Expert Use of Representation

This level of representational competence refers to students' creation of multiple representations and using them to interpret, identify or explain an underlying phenomenon like an expert in the corresponding field of study. To have an expert level of representational competence in any field, one must be able to create multiple representations that connect concepts from multiple systems. Additionally, experts use evidence to support the representations they create and can accurately describe the meaning behind the representations. For instance, Bruce generated a representation that used different scales (i.e. double-helix to histones to chromosome to entire DNA). As he was drawing he stated, "this baseline is pretty much the genome itself, and then within each section you have forward and reverse genes because it can go both ways. And so, it would – the annotation kinda selects a certain – creates – identifies a certain gene within that genome. And then with the annotation it can give a whole bunch of different facts about it" (Fig. 16). Bruce then goes on to describe what annotated genomes are and how scientists use them,

> More or less look at the different genes within that genome and identify a certain sequence as a particular gene. And then, I give the characteristics of that gene such as its functionality, its length, a whole bunch of data about it.... Being able to look at the functionality of different phages or whatever you're looking at and being able to identify what it does and from that you can use it to further – do further research into developing new phages or whatever and just identifying functionalities and stuff like that and hopefully gain more knowledge for future usage.

Bruce is able to describe the location and use of annotations, through multiple representations, exhibiting an expert level of representational competence with annotated genomes.

During Jack's post-instruction interview, he first drew a graph that represented what he saw in the program GeneMarkTB and then he drew a series of base-pairs linked together in boxes (Fig. 17). Underneath the boxes he wrote "Gene 1." As he was drawing the different representations he described things to consider when annotating a genome,

Fig. 17 Jack's post
representation

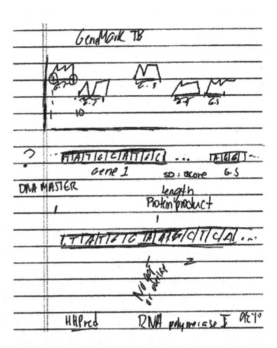

So first off, there are several interfaces in the internet and software that you run your entire
sequence through … this one's called GeneMark; it just tells you what the probability is that
there is a gene, it just draws a graphical interpretation of where there should be, several
genes within a specific sequence. So, what I drew here is that the program, the program
draws some lines above it or doesn't draw anything whatsoever. If it draws like a lot of lines
like the length of the line is, that just tells you that there's a strong possibility that that length
is the genes coding potential. According to those sequences it makes a draft of how long it
should be. So if you; if we continue down all the genome like this is how annotated genome
looks like.

As Jack is describing this, he is drawing the different images and using them in his
description like a scientist would. The use of authentic bioinformatics software and
engaging students in authentic scientific practice is helping students change and
become more like a scientist in their thinking as revealed by the following representations and discussion.

Changes in Students' Visualizations

Participants in the CURE had the opportunity to complete both a questionnaire and
an interview about annotated genomes. Questionnaires were completed before and
after the CURE experience while interviews were conducted between and at the end
of the two semester CURE. We compared the changes in students' levels of

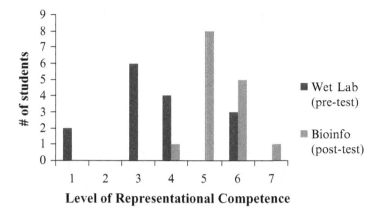

Fig. 18 Changes in student's level of representational competence before and after participation in the CURE

representational competence with annotated genomes pre and post to determine if there were any changes after participation in the CURE. At the beginning of the semester the majority of students' representations fell into Level 3: Simplified Use (40%) with no students' representations existing at Level 7: Expert Use.

Representational Competence Increased after Participation in CURE

We examined the changes in representational competence that occurred for students participating in both semesters of the CURE. We found a significant increase ($p = 0.0002$) in student's representational competence after participation in the two-course series (Fig. 18).

We examined the quantitative change in students' level of representational competence pre and post CURE. Students either remained at the same level (0, no change) or increased in their level of representational competence ranging from a positive gain of 1 to 5 (Fig. 19). The analysis revealed that no students decreased in their levels of representational competence after participation in the CURE (0 students had a negative value). Only three students remained at the same level (0 change) while the remaining students increased: three students increased one level, six students increased two levels, two students increased four levels, and one student increased five levels pre and post CURE.

Representative Examples of Students' Changes of Representational Competence

Annika and Jack were the two most drastic changes after participation in the CURE. Both students increased four levels of representational competence (Fig. 20). Annika started at Level 1 by stating that she did not know what an annotated genome

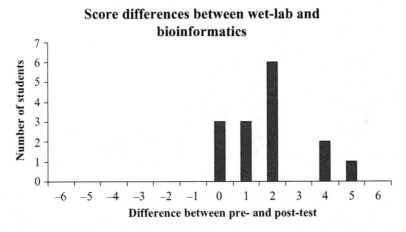

Fig. 19 Score differences between pre-test (beginning of wet-lab semester) and post-test (end of bioinformatics semester)

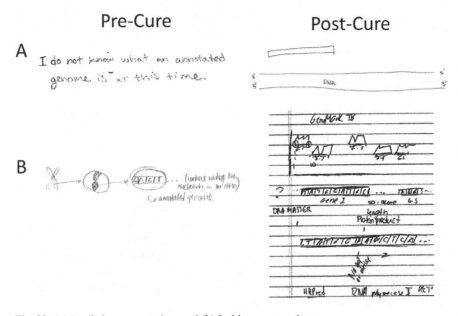

Fig. 20 (a) Annika's representations and (b) Jack's representations

was at the beginning of the course. At the end of the second semester of the CURE, Annika reached a representational competence of Level 5, while Jack progressed from Level 3 to Level 7.

Conclusions and Implications

It has been documented that students struggle to learn concepts of genes and genomes in genetics (e.g., Driver et al. 1994; Lewis et al. 2000). The definition of a gene has been modified over time to reflect increases in scientific knowledge and changes in technology and is still debated by genomics research scientists, biological scientific philosophers, and educators (Dikmenli et al. 2011; Gericke and Hagberg 2007; Griffiths and Neumann-Held 1999; Knight; Stotz et al. 2004). Some have posed that a scientist's view of a gene is grounded in their experimental perspective: developmental, evolutionary, or molecular (e.g., Fogle 2001; Rheinberger and Muller-Wille 2008; Stotz et al. 2004; Waters 1994). In addition, the definition of a gene has been impacted by the large amounts of data obtained from genomic sequencing projects and the use of technology and computational tools for annotation, or making sense of the genome sequences through annotation. The change in technology and the abundance of data has changed the way that we think about genomes and genomic information. As new technologies improve, new experimental evidence is provided and scientists are able to examine genes in ever greater detail, the definition of a gene by scientists has continued to be elusive and a source of debate, but consistent is the inclusion of function with the sequence and the application of the definition to be used by the technology for genome annotation projects (Baetu 2012; Gerstein et al. 2007).

We found the Halverson and Friedrichsen (2013) framework for representational competence useful within another context area of biology beyond tree thinking. The framework allowed us to identify and note changes in students thinking about genomes, and while we minimally modified the descriptions based on the new content, the levels still held. This allows for a way to examine the impact of CURE experiences on students' representational competence by looking at both content and process components. In order to annotate a genome, a student must understand the scientific content from the biological system and develop a process understanding that requires the use of technology to evaluate multiple pieces of evidence gathered from databases and gleaned from multiple representations displayed by bioinformatics software. Evaluation of the evidence and interpretation of the representations relies on building connections back to the implications in the overall biological system. An understanding of function is critical in addition to the sequence and an understanding of how technology is used to gather evidence and provide representations that help lead to new discovery. We found this distinction among the levels of representational competence for the students. An understanding of both content and process, including the process of using technology to uncover new understanding and discover new knowledge to build deeper connections among the components within the biological system. At lower levels, the students demonstrate an understanding of the surface level (the nucleotide sequence), then they begin to discuss and draw the basics of how to use the technology (to define/annotate/assign meaning to portions and pieces and parts of the nucleotide sequence) and then they understand function (and gaps in knowledge, based upon limitations in the evidence or technology that is used to build that knowledge of function and the resulting putative gene annotation) and build deeper connections to the overall biological system—at the highest levels of representational competence students connect all three and support their understanding with rich descriptions and multiple representations.

The lack of research on the role that representations play in learning biology, particularly in introductory courses, makes implementation and refinement of representations difficult in biology (Griffard 2013). Research has shown:

> Students' construction of knowledge in biology is closely related to an ability to translate across and between Multiple External Representations represented at various levels of organization. Promoting skill-based translation practices for advancing our students' biological understanding should be viewed as a key enterprise of modern biology teaching (Schonborn & Bogeholz 2013, p. 126).

Students' use of representations helps us capture changes in their process skills, specifically as they relate to working with representations. It is important to realize that students' representational competence is impacted and changes from classroom experiences (e.g. Stieff et al. 2005). Given that representational competence is something that can change and we know that it is important for expertise in science, it is critical that we pay attention to it. This is true for both the representations we use to communicate and teach science within the classroom and also the tasks that we ask students to undertake as they work with representations, including in the technology rich environments of modern science/biology. Representations and how they are communicated can be used to reveal changes in thinking. Our study suggests that providing an authentic CURE experience with genome annotation can provide a way for students to practice working among Multiple External Representations and further achieve a higher level of representational competence. More specifically, in early levels of competence, students noted links to terminology but not to the process or the broader conceptual understanding of thinking about the biological system. Perhaps students plateaued initially around level 3 and 4 because they did not have an opportunity to work with multiple representations of the biological content within the system and it may be that a bioinformatics course within a CURE can help students build deeper conceptual understanding and scientific thinking. Future work is needed to explore the specific impact of the bioinformatics course, the ability of students to solve problems with representations in more detail and the exploration of questions appropriate for other content areas in biology.

Acknowledgments This research and interdisciplinary collaboration was supported in part by a visionary Grant from the Gordon Research Conference on Visualization in Science and Education (2009), the National Institute of General Medical Sciences from the National Institutes of Health, Howard Hughes Medical Institute, the University Grants Development Program and University, Biotechnology Innovation and Regulatory Science Center, Polytechnic Institute, Department of

Agricultural and Biological Engineering, Department of Technology Leadership and Innovation, Purdue University. We would also like to thank the Authors Research labs for all of their work to make this possible.

References

Anderson, T. R., Schönborn, K. J., du Plessis, L., Gupthar, A. S., Hull, T. L. (2012). Identifying and developing students' ability to reason with concepts and representations in biology. In D.F. Treagust & C. Tsui (Eds.), *Multiple representations in biological education.* (pp. 19–38). doi: https://doi.org/10.1007/978-94-007-4192-8_2

Baetu, T. M. (2012). Genomic programs as mechanism schemas: A non-reductionist interpretation. *British Journal for the Philosophy of Science, 63,* 649–671.

Baum, D. A., Smith, S. D., & Donovan, S. S. S. (2005). The tree-thinking challenge. *Science, 310,* 979–980.

Chi, M. T. H., Feltovich, P. J., & Glasner, R. (1981). Categorization and representation of physics problems by experts and novices. *Cognitive Science, 5,* 121–152.

Cuoco, A. A., & Curcio, F. R. (2001). *The roles of representation in school mathematics.* National Council of Teachers.

Dees, J., Momsen, J. L., Niemi, J., & Montplaisir, L. (2014). Student interpretations of phylogenetic trees in an introductory biology course. *CBE-Life Sciences Education, 13*(4), 666–676.

Dikmenli, M., Cardak, O., & Kiray, S.A. (2011). Science student teachers' ideas of the concept 'gene'. In annual meeting of the 3rd world conference on educational sciences, Istanbul, Turkey.

Driver, R., Squires, A., Rushworth, P., & Woods-Robinson, V. (1994). *Making sense of secondary science: Research into children's ideas.* London: Routledge.

Ferk, B., Vrtacnik, M., Blejec, A., & Gril, A. (2003). Students understanding of molecular structure representations. *International Journal of Science Education, 25,* 1227–1245.

Fogle, T. (2001). The dissolution of protein coding genes in molecular biology. In P. Beurton, R. Falk, & H.-J. Rheinberger (Eds.), *The concept of the gene in development and evolution.* Cambridge: Cambridge University Press.

Gericke, N. M., & Hagberg, M. (2007). Definition of historical models of gene function and their relation to student's understanding of genetics. *Science & Education, 16,* 849–881.

Gerstein, M. B., Bruce, C., Rozowsky, J. S., Zheng, D., Du, J., Korbel, J. O., Emanuelsson, O., Zhang, Z. D., Weissman, S., & Snyder, M. (2007). What is a gene, post-ENCODE? History and updated definition. *Genome Research, 17,* 669–681.

Gilbert, J. K. (2005a). Visualization: A metacognitive skill in science and science education. In J. K. Gilbert (Ed.), *Visualization in science education* (pp. 9–27). Dordrecht: Springer.

Gilbert, J. K. (2005b). *Visualizations in science education (Vol 1).* Dordrecht: Springer.

Gilbert, J. K., & Treagust, D. (Eds.). (2009). *Multiple representations in chemical education (Vol. 4).* Dordrecht: Springer.

Griffard, P. B. (2013). Deconstructing and decoding complex process diagrams in university biology. In D. Treagust & C.-Y. Tsui (Eds.), *Multiple representations in biology education (chapter 10).* Dordrecht: Springer.

Griffiths, P. E., & Neumann-Held, E. M. (1999). The many faces of the gene. *Bioscience, 49*(8), 656–662.

Halverson, K. L. (2010). Using pipe cleaners to bring the tree of life to life. *The American Biology Teacher, 72*(4), 223–224.

Halverson, K. L. (2011). Improving tree-thinking one learnable skill at a time. *Evolution: Education and Outreach, 4*(1), 95–106.

Halverson, K. L., & Friedrichsen, P. (2013). Learning tree thinking: Developing a new framework of representational competence. In *Multiple. representations in biological education* (pp. 185–201). Dordrecht: Springer Netherlands.

Halverson, K. L., Pires, C. J., & Abell, S. K. (2011). Exploring the complexity of tree thinking expertise in an undergraduate systematics course. *Science Education, 95*(5), 794–823.

Harle, M., & Towns, M. (2010). A review of spatial ability literature, its connection to chemistry, and implications for instruction. *Journal of Chemical Education, 88*(3), 351–360.

Harle, M., & Towns, M. H. (2012). Students' understanding of external representations of the potassium ion channel protein part II: Structure–function relationships and fragmented knowledge. *Biochemistry and Molecular Biology Education, 40*(6), 357–363.

Harle, M., & Towns, M. H. (2013). Students' understanding of primary and secondary protein structure: Drawing secondary protein structure reveals student understanding better than simple recognition of structures. *Biochemistry and Molecular Biology Education, 41*(6), 369–376.

Harrison, M., Dunbar, D., Ratmansky, L., Boyd, K., & Lopatto, D. (2011). Classroom-based science research at the introductory level: Changes in career choices and attitude. *CBE-Life Sciences Education, 10*(3), 279–286.

Jordan, T. C., Burnett, S. H., Carson, S., Caruso, S. M., Clase, K., DeJong, R. J., et al. (2014). A broadly implementable research course in phage discovery and genomics for first-year undergraduate students. *MBio, 5*(1), e01051–e01013.

Kozma, R., & Russell, J. (2005). Students becoming chemists: Developing representational competence. In J. K. Gilbert (Ed.), *Visualization in science education* (pp. 121–145). Dordrecht: Springer.

Kozma, R., & Russell, J. (2007). Modelling students becoming chemists: Developing representational competence. In J. K. Gilbert (Ed.), *Visualization in science education* (pp. 147–168). Dordrecht: Springer.

Lewis, J., Leach, J., & Wood-Robinson, C. (2000). All in the genes? – Young people's understanding of the nature of genes. *Journal of Biological Education, 34*, 74–79.

Mathewson, J. H. (1999). Visual-spatial thinking: An aspect of science overlooked by educators. *Science Education, 83*, 33–54.

Matuk, C. (2007). Images of evolution. *Journal of Biocommunication, 33*(3), E54–E61.

Meyer, M. R. (2001). Representation in realistic mathematics education. In A. A. Cuoco (Ed.), *The roles of representation in school mathematics (2001 yearbook)* (pp. 238–250). Reston: National Council of Teachers in Mathematics.

National Research Council. (1996). National science education standards. National science education standards: National Academy Press.

Patrick, M. D., Carter, G., & Wiebe, E. N. (2005). Visual representations of DNA replication: Middle grades students' perceptions and interpretations. *Journal of Science Education and Technology, 14*, 353–365.

Peterson, M. P. (1994). Cognitive issues in cartographic visualization. In A. M. MacEachren & D. R. F. Taylor (Eds.), *Visualization in modern cartography* (pp. 27–43). Oxford: Pergamon.

Pruitt, K. D., Tatusova, T., Brown, G. R., & Maglott, D. R., (2011). *NCBI reference sequences (RefSeq): Current status, new features and genome annotation policy.* Nucleic Acids Research, Advance Access, 1–6.

Rheinberger, H.-J., & Muller-Wille, S. (2008). Gene concepts. In S. Sahotra & A. Plutynski (Eds.), *A companion to the philosophy of biology* (pp. 3–21). Oxford: Blackwell Publishing.

Roth, W.-M., Bowen, G. M., & McGinn, M. K. (1999). Differences in graph-related practices between high school biology textbooks and scientific ecology journals. *Journal of Research in Science Teaching, 36*, 977–1019.

Rutherford, J. F., & Ahlgren, A. (1990). *Science for all Americans.* New York: Oxford University Press.

Schönborn, K. J., & Bögeholz, S. (2013). Experts' views on translation across multiple external representations in acquiring biological knowledge about ecology, genetics, and evolution. *In Multiple representations in biological education* (p. 126). Springer Netherlands.

Shaer, O., Kol, G., Strait, M., Fan, C., Grevet, C., & Elfenbein, S. (2010). G-nome surfer: A tabletop interface for collaborative exploration of genomic data. In *Proceedings of human factors in computing systems (1427–1436).* New York: ACM Press.

Shaer, O., Strait, M., Valdes, C., Wang, H., Fend, T., Lintz, M., Ferreirae, M., Grote, C., Tempel, K., & Liu, S. (2012). The design, development, and deployment of a tabletop interface for collaborative exploration of genomic data. *International Journal of Human-Computer Studies, 70*(10), 746–764.

Shepard, R. (1988). The imagination of the scientist. In K. Egan & D. Nadaner (Eds.), *Imagination and education* (pp. 153–185). New York: Teachers' College Press.

Singer, S. R., Nielsen, N. R., & Schweingruber, H. A. (Eds.). (2014). *Discipline-based education research: Understanding and improving learning in undergraduate science and engineering.* Washington, D.C.: National Academies Press.

Sterk, P., Kersey, P. J., & Apweiler, R. (2006). Genome reviews: Standardizing content and representation of information about complete genomes. *OMICS: A Journal of Integrative Biology, 10*(2), 114–118.

Stieff, M., Bateman, R. C., Jr., & Uttal, D. H. (2005). Teaching and learning with three dimensional representations. In J. K. Gilbert (Ed.), *Visualization in science education* (pp. 93–118). Netherlands: Springer.

Stotz, K., Griffiths, P. E., & Knight, R. (2004). How biologists conceptualize genes: An empirical study. *Studies in History and Philosophy of Science Part C., 35*(4), 647–673.

Takayama, K. (2005). Visualizing the science of genomics. In J. K. Gilbert (Ed.), *Visualization in science education* (pp. 217–252). Netherlands: Springer.

Trumbo, J. (1999). Visual literacy and science communication. *Science Communication, 20*(4), 409–425.

Tytler, R., Prain, V., Hubber, P., & Waldrip, B. (Eds.). (2013). Constructing representations to learn in science.New York:Springer Science & Business Media.

Waldrip, B., & Prain, V. (2012). Learning from and through representations in science. In B. J. Fraser, K. Tobin, & C. J. McRobbie (Eds.), *Second international handbook of science education* (pp. 145–155). Dordrecht: Springer.

Waters, C. K. (1994). Genes made molecular. *Philosophy of Science, 61*, 163–185.

Won, M., Yoon, H., & Treagust, D. F. (2014). Students learning strategies with multiple representations: Explanations of the human breathing mechanism. *Science Education, 98*(5), 840–866.

Yore, L. D., & Hand, B. (2010). Epilogue: Plotting a research agenda for multiple representations, multiple modality, and multimodal representational competency. *Research in Science Education, 40*(1), 93–101.

Part III
The Assessment and Attainment of Representational Competence

Using Gesture Analysis to Assess Students' Developing Representational Competence

Matthew E. Lira and Mike Stieff

Introduction

Many scientists whose discoveries impact today's society made those discoveries by inventing external representations to support their thinking and communicating (Crick & Watson 1954). External representations play a role so critical in science many educators strive to understand, develop, and assess how science students use representations to learn. Students' use of representations to think and communicate refers to a broad set of skills called Representational Competence. More specific, Representational Competence (RC) refers to a constellation of skills that involve interpreting, generating, and manipulating external representations to support learning, problem solving, and communicating in STEM fields (diSessa 2004; Kozma and Russell 2005).

In STEM disciplines, students must learn to coordinate manifold concepts that vary across levels of organization. Each level of organization is described by a host of representations. Science educators often task students with learning to coordinate the lower-level organization of a system with its higher-level aggregate properties. For instance, chemistry students must learn to recognize that bubbles evolve from a chemical reaction when two sub-microscopic reactants form a gaseous product that manifests at the macroscopic level. To learn about this phenomenon, students must interpret chemical formulas, space-filling diagrams, and plots of concentration.

M. E. Lira (✉)
The University of Iowa, Iowa, IA, USA
e-mail: matthew-lira@uiowa.edu

M. Stieff
University of Illinois at Chicago, Chicago, IL, USA
e-mail: mstieff@uic.edu

© Springer International Publishing AG, part of Springer Nature 2018
K. L. Daniel (ed.), *Towards a Framework for Representational Competence in Science Education*, Models and Modeling in Science Education 11,
https://doi.org/10.1007/978-3-319-89945-9_10

Learning in STEM thus demands "multi-level thought" (Johnstone 1991). Students must learn to interpret representations that correspond to multiple levels of organization and understand how the representations relate.

The learning challenge we describe above has been investigated extensively through the lens of RC. Many investigations assess differences between experts and novices or novices before and after instruction. Such investigations often produce an alarming picture of students' RC (Ainsworth 2006). Although our knowledge about RC continues to grow, we still know little about how students' RC develops. Current models of the construct suggest that students' content knowledge predicts their use of representations but not visa versa (Nitz et al. 2014). Such findings pose a threat to the validity of the construct. Although we have measures to assess RC independent from content knowledge, RC appears epiphenomenal.

We propose an alternative. We interpret students' difficulties with achieving mastery over RC as a sign that we should re-evaluate the grain size of our measures and consider alternative techniques to assess how students develop RC. Many other investigations leverage microgenetic methods to detecting fine-grained changes in students' naïve and developing RC (Azevedo 2000; Hammer et al. 1991; Sherin 2000). We applaud these efforts to understand developing RC but argue that because these investigations use student-generated inscriptions (e.g. graphs) in communities of practice, we may still miss mechanisms of knowledge transition in individuals. Drawings, for instance, constitute the end product of process that demands many cognitive processes (Jolley 2010). These cognitive processes prove difficult to identify when assessing the end product of the drawing itself. Moreover, we are less likely to understand individual mechanisms of knowledge transition if the unit of analysis is the community rather than the individual. To assess the development of RC in an individual we need more fine-grained measures that detect developing RC in ways that inscription tasks do not.

Gesture analysis provides one such assessment technique. A body of evidence shows that gesture analysis provides robust insight into mechanisms of knowledge transition. We argue that gesture analysis constitutes a formative assessment technique for measuring STEM students' developing RC. Gesture analysis provides insight into intermediary knowledge states and learning mechanisms among populations with ages that range from 5 years old (Ehrlich et al. 2006) to adults (Garber and Goldin-Meadow 2002) and from domains as diverse as word learning (Iverson and Goldin-Meadow 2005) and organic chemistry (Stieff 2011). The critical affordance of gesture to assess the development of RC in STEM fields lies in its capacity to represent spatial and dynamic information.

In this chapter we introduce a framework for using gesture to assess the development of RC. We apply the framework in a pilot investigation where we executed a gesture analysis technique to assess how students' develop RC when reasoning about a biological system: the resting membrane potential. We chose this concept because it is a biological phenomenon modeled with concepts from physics and mathematics. The membrane potential is quantified as a voltage and traditional textbooks and lecture materials coordinate the concept with disciplinary representations that include graphs, equations, and circuit diagrams. In learning this concept, students are tasked with understanding the relation between multiple external disciplinary representations to understand models of cells in physiology. We find that students' gestures reveal their developing RC in this discipline by (1) describing entities and processes, (2) identifying critical features of disciplinary representations, and (3) relating external representations. These findings motivate further work to determine the representativeness of these episodes and further efforts to examine the validity of gesture analysis by assessing students' gestures alongside more vetted measures of RC, such as constructing, interpreting, or translating between multiple disciplinary representations.

Assessing Representational Competence in STEM Disciplines

Prior Approaches to Assessing Representational Competence

RC is a defining characteristic of expertise in science. Scientists use external representations to think through the problems they face in research (Crick and Watson 1954); scientists construct and interpret representations not as an ancillary activity but as an integral component in their pursuit to understand phenomena and make discoveries. In many instances humans cannot perceive the entities and processes under investigation because they exist and occur at scales beyond our senses and life spans. External representations improve the way scientist understand problems and new representations unlock new avenues of research and discovery. External representations serve the needs of both the individual and the community. Some have argued that science simply cannot be done without external representations that permit the community to develop a shared understanding (Latour 1986). The history of the atom serves as a prominent example. Without transforming mental models of the atom into external representations, scientists would have struggled to argue about the assumptions of the competing models and the necessary experiments to test the propositions of each respective model.

The central role that external representations play in scientific practice motivate STEM educators to help students develop RC to improve their understanding of disciplinary concepts and the nature of science. RC remains a generative construct for identifying barriers to conceptual change and successful problem solving in science education. Working across STEM disciplines and using diverse methodologies, investigators continue to explore how experts and novices construct, translate, and relate multiple external representations as well as invent novel ones. Prior approaches to assessing students' RC have provided valuable information that has identified the learning challenges students face and the instructional targets that educators should address. In efforts to assess students' RC, many investigations leverage the suite of skills defined by Kozma and Russell (2005):

> The ability to use representations to describe observable chemical phenomena in terms of underlying molecular entities and processes.

The ability to generate or select a representation and explain why it is appropriate for a particular purpose.

The ability to use words to identify and analyze features of a particular representation (such as a peak on a coordinate graph) and patterns of features (such as the behavior of molecules in an animation).

The ability to describe how different representations might say the same thing in different ways and explain how one representation might say something different or something that cannot be said with another.

The ability to make connections across different representations, to map features of one type of representation onto those of another (such as mapping a peak of a graph onto a structural diagram), and to explain the relationship between them.

The ability to take the epistemological position that representations correspond to but are distinct from the phenomena that are observed.

The ability to use representations and their features in social situations as evidence to support claims, draw inferences, and make predictions about observable chemical phenomena (p. 132).

These definitions for RC demonstrate their intellectual contribution by the volume of research they generated. We recognize the merit in the skills described above and encourage educators to continue to develop these skills in students. Educators must work strategically to align instruction with assessments and in turn address the manifold learning challenges experienced by students. Students often fail to use representations to their advantage because they focus on superficial features between representations as opposed to conceptual relations (Chi et al. 1981), ignore useful representations during problem solving (Stieff et al. 2011), treat representations literally as if they were the referent (Uttal and O'Doherty 2008), and make inaccurate translations and interpretations (McDermott et al. 1987; Shah and Hoeffner 2002). The chasm between novices and experts raises the question of how students successfully achieve RC. Assessing the development of this transition remains a methodological challenge as much as a theoretical one: it is easy to identify when students demonstrate RC (or the lack thereof), but it is more difficult to assess the development of RC.

One strategy for overcoming the limitations of prior RC assessments involves assessing students' verbalizations and visual attention while they interpret and relate external representations. This method of assessment enables educators to detect when students use representations to develop understanding as opposed to using their understanding to interpret or generate representations. An investigation in physics education provides an illustrative example (Parnafes 2007). Parnafes specified mechanisms of knowledge transition by determining the features of representations that students attended to in their utterances, gaze, and deictic gestures. This technique illustrated how students' intuitions about the meanings of representations changed as they interacted with the representations. For instance, Parnafes attends to students' content knowledge by demonstrating that students hold multiple meanings for one concept. Thus, "fast" might mean both high velocity and high frequency despite the semantic distinction drawn in physics. When students think and learn with multiple external representations of velocity and frequency, these representations facilitate learning because they transform ephemeral temporal events into permanent spatial ones that make differences salient to students.

The critical contribution of Parnafes' work comes from her detection of learning mechanisms vis-à-vis disciplinary representations. Before students possess the skill to articulate disciplinary concepts in words or in inscriptions, students develop the skill to coordinate perceptual foci with their intuitions. She noted that students first detect patterns in external representations (a component of RC) and only later identify the patterns' correspondence to physical entities and processes. For instance, students notice that two graphs of sine waves have the same period before they understand that two oscillating pendulums maintain a constant frequency independent of instantaneous velocity. Evidence for students' developing RC appears in their interactions with representations before students attain robust conceptual understandings of disciplinary concepts. This investigation illustrates how sensitive measures of RC improve our understanding of how students use representations to modify their intuitions.

The integral relation between students' knowledge and RC presents opportunities but also challenges to RC assessments. To assess students' content knowledge independent from their developing RC, investigators measure students' accuracy on multiple-choice text and representational transformation tasks respectively (Nitz et al. 2014). Students' *content* knowledge predicts their RC but not visa-versa. This finding poses a threat to the validity of the construct. Although RC and content knowledge can be measured independently, RC is rendered epiphenomenal—knowledge transition occurs through unseen cognitive mechanisms and then later manifests itself in students' uses of representations. Such a picture at first seems contradictory to the Parnafes (2007) investigation that illustrated how students use representations to change their understanding of disciplinary concepts. These results are not mutually exclusive. Parnafes characterized students' *intuitive* knowledge as opposed to their content knowledge per se. Students' intuitions about disciplinary concepts and representations might provide more sensitive measures to witness how RC emerges before conceptual change.

Eye tracking provides one such precise and non-intrusive measure. By using eye tracking investigators gain rigorous and sensitive measurements of how students examine features of external representations and use these representations to solve problems. Eye tracking investigations, however, provide a picture similar to other efforts that assess the relation between students' content knowledge and RC: students with high content knowledge use disciplinary representations to learn and problem solve more than students with low content knowledge (Cook et al. 2008). For instance, chemistry students with high content knowledge use unfamiliar representations to solve problems significantly more than students judged to possess low content knowledge (Hinze et al. 2013). Findings from eye tracking investigations therefore again render RC epiphenomenal if we accept these measures. We offer an alternative view. We suggest that because these investigations present students with formal disciplinary inscriptions, they demand significant requisite content knowledge from students. If we aim to assess how students' use their developing RC to learn and communicate, we must measure students' intuitive notions about representation in intuitive modes of assessment.

Using Gesture to Assess the Development of RC

Following the theoretical and empirical work of David McNeill (1992), we espouse the assumption that speech and gesture constitute one neurocognitive system that shares one idea unit during each utterance. The spoken word carries the linguistic component of the idea and gesture carries the visual component. We make this point to highlight that when speaking and gesturing the speaker automatically brings visual representations into thought and connects those representations to linguistic descriptions. Educators bemoan the recurring challenge that students face when translating disciplinary representations into words (Kozma and Russell 1997). Gesture production during speech provides one avenue for students to begin to build connections between nascent encodings of visual representations and their corresponding explicit linguistic descriptions. Assessing gestures should shed light on students' developing RC when they struggle to make their knowledge explicit in speech and other modes.

Gesture production involves a tight neurological coupling to speech production in development and disease (McNeill 1985). In pathology, speech and gesture production deteriorate in parallel for patients suffering from Broca's area aphasia (Pedelty 1985). Furthermore, early on during paraphasia, patients experience transition states whereby their gestures represent information relevant to the referent even when their words do not. During development, children's gesture and speech production manifest in tandem. Gesture production in infancy and early childhood predicts later word learning (Iverson and Goldin-Meadow 2005) and even later narrative production and reading comprehension (Demir 2009). These findings demonstrate that the cognition supporting speech and gesture production share a common computational stage and thus support the argument that gesture holds equal status to that of speech when assessing a speakers' knowledge.

Not all gestures are equally useful for assessing RC. By gesture we refer to the movements of the arms and hands that typically co-occur with speech. Although no single taxonomy is agreed upon, researchers generally agree on a few broad classes. **Deictic** gestures constitute the first class and refer to gestures that involve pointing to objects, inscriptions, or locations whether real or imagined. These gestures often co-occur when speakers communicate direction. **Beat** gestures constitute a second class and refer to the flicks of the hands that occur rhythmically during longer utterances such as narratives. Speakers often use beat gestures to segment speech or organize ideas. **Emphatic** gestures constitute a third class and refer to swift or abrupt motions of the hands and arms that co-occur with changes in the pitch, prosody, or volume of the voice to call the interlocutor's attention to the stress placed on particular utterances within a sentence.

Representational, or iconic, gestures constitute the last class and refer to gestures that represent the imagistic content of verbal referents. Thus, these gestures represent visual, spatial, and temporal information. When speakers construct representational gestures their hands or arms often assume shapes that bear realistic similarity to the entities and processes they represent. Extending the index and middle finger together to represent the ears of a dog or rabbit provide a common example. Contemporary approaches caution against forcing gestures into categories with rigid boundaries—natural gestures belong to many classes at once. Although each class of gesture has a functional role in cognition and communication, we expect representational gestures to prove especially useful for gaining insight into a students' developing RC. We select representational gestures because they provide the clearest connection to RC given that they represent entities and processes similar to that of disciplinary representations.

Representational gestures hold specific affordances that support the demands of assessing student knowledge and skill in STEM. Many concepts in STEM involve spatio-temporal information. This information is easier to communicate in gesture than in speech. A critical mass of research illustrates that people use gestures to support spatial thinking and communicating (Alibali 2005; Chu and Kita 2011; Garber and Goldin-Meadow 2002). Students' gestures provide insight into mechanisms of knowledge transition and because gesture constitutes a natural mode of communication instructors do not need to teach students to produce gesture. Just as formal disciplinary representations emphasize and deemphasize particular aspects of disciplinary concepts to support thinking and communicating, gestures provide students the opportunity to construct or interpret schematic representations that depict critical features of their referent.

For example, the representational gestures that students produce during retrospective think aloud protocols provide insight into students' developing mathematical understanding (Perry et al. 1988). In a seminal report, these investigators demonstrated that among students who failed to solve mathematical equivalence problems (e.g. 2 + 3 = __ + 1), a subset of students produced gestures that grouped mathematical terms—suggesting that they noticed the addend but failed to use the information to solve the problem. Other students who also failed to solve the problems did not produce such gestures. Students who produced the grouping gesture benefited more from instruction than students who did not produce the gesture.

Results similar to the one above demonstrate that representational gesture predicts readiness to learn. Gestures reveal visual-spatial and spatio-temporal knowledge that may not be accessible to verbal modes. By attending to peoples' gestures we can predict their strategy use, performance, and receptiveness to instruction in tasks that range from Piagetian conservation (Church and Goldin-Meadow 1986) to gear-rotation (Schwartz and Black 1996). The fact that gesture plays a generative role across ages and domains highlights the robust role of gesture in assessing student learning and supports our efforts to use gesture analysis to assess students' developing RC.

Further rationale for studying students' gestures comes from work that demonstrates that students use gestures to communicate in STEM disciplines that include biology (Srivastava and Ramadas 2013), geology (Kastens et al. 2008), chemistry (Flood et al. 2014), mathematics (Alibali and Nathan 2012) and physics (Scherr 2008). For instance, physics students' gestures serve a bridging role as they transition from producing novice-like to expert-like discourse (Roth 2000). Before students comprehend and appropriate disciplinary formalisms such as operational definitions and disciplinary diagrams, students use gestures to supplement and complement their verbal descriptions of laboratory observations. Thus, the gestures students produce during authentic scientific discourse in classrooms reveals their developing knowledge in ways similar to the results of laboratory investigations.

Roth's results echo other efforts that document the generative role of gesture in scientific communities. For instance, an ethnographic investigation that took place in a biochemistry laboratory revealed that researchers use gesture as a tentative model that described the protein they were studying (Becvar et al. 2005). Professional scientists using gesture to represent scientific concepts lend credence to our assertion that gesture constitutes an authentic and generative mode of representation in STEM. Thus, the gestures that students produce during verbal communication may supplement or augment our understanding of their developing RC.

A Framework for Assessing RC with Gesture

Representational gestures hold the potential to support students' efforts to communicate their content knowledge during mechanistic reasoning. To communicate in many STEM disciplines, scientists must describe unseen entities and processes (Johnstone 1991). As Kozma and Russell (2005) claim, "representations—such as written or drawn symbols, iconic [or representational] gestures or diagrams— "stand for" or "refer to" other objects or situations," (p. 130). Just as professional biochemists use gesture to communicate information regarding protein conformation and conformational changes, STEM students can use gesture to represent any unseen entities and processes broadly. Because gestures co-occur naturally with speech, students' words should coordinate with their contemporaneous gestures. Furthermore, because many unseen entities and processes at lower levels of organization result in emergent phenomena at higher levels, mechanistic reasoning requires students to coordinate multiple levels of organization and thus the corresponding representations that describe these levels. Given the fluid nature of gesture, students' gestures should transition from one form to the next while they reason. As students verbalize ideas with individual words in speech and represent them in gesture, their larger verbal explanation should provide a supporting narrative that relates the different representational gestures. From this analysis of the task demands of mechanistic reasoning, we predicted that students would demonstrate their developing RC on three of the skill constructs described by Kozma and Russell:

> The ability to use representations to describe observable [chemical] phenomena in terms of underlying [molecular] entities and processes.

The ability to use words to identify and analyze features of a particular representation (such as a peak on a coordinate graph) and patterns of features (such as the behavior of molecules in an animation).

The ability to make connections across different representations, to map features of one type of representation onto those of another (such as mapping a peak of a graph onto a structural diagram), and to explain the relationship between them (p. 130).

These three skill constructs aligned with our efforts to assess students' developing RC by identifying representational skills relevant to the task demands of mechanistic reasoning. Regarding the remaining skill constructs, we selected against them because they would require students to call explicit attention to their gestures. For instance, for a student to explain why their representational gesture is appropriate they would have to refer verbally to their gesture. Although exceptions exist, when people gesture their gesture production occurs below the level of conscious awareness (McNeill 2005). Because we did not anticipate students to call explicit attention to their gestures, we eliminated skill constructs that would require students to do so.

In contrast, we see principled rationale for finding evidence of the three remaining skills in mechanistic reasoning tasks. Regarding the first skill, verbal descriptions often fail to communicate spatio-dynamic information. When people tell tales about fishing excursions, they don't just say, "I caught a big one." They say, "I caught one *this* big!" and supplement the demonstrative term with a gesture that represents the size of the fish. Similarly, when students describe entities, their properties, and the processes involved in a mechanism, their verbal descriptions should fail to communicate spatio-dynamic information. Students' representational gestures should describe the missing but critical information.

Regarding the second skill, we discussed earlier how students often fail to distinguish between the different meanings of related science terms (Parnafes 2007). When students struggle to verbally characterize a concept, their gestures can clarify the meaning of the referent. Communicating mechanistic reasoning places gesture in a higher position still because students must communicate how entities interact to cause changes in the properties of a system. Gesture analysis reveals that children produce gestures that represent causality before causal relations appear in speech (Göksun et al. 2010). If given the opportunity to express their intuitions, student's verbalizations and co-occurring representational gestures should supplement one another.

Regarding the third skill, relating multiple external representations poses a number of cognitive challenges to students, from selecting and attending to relevant features while ignoring others, to retrieving and organizing ideas into a coherent explanation. Students are learning how to accomplish this task but before they master it, their gestures should illustrate their developing RC by showing how they re-represent features of disciplinary representations in series and describe them in sequence. With our rationale presented, we offer below a set of conjectures for how gesture analyses might help educators assess students' developing RC:

When students struggle to construct disciplinary inscriptions or verbal descriptions of entities and processes, gesture analysis reveals their efforts because gesture affords spatio-dynamic representation.

> When students struggle to coordinate words or text with other symbols and icons in disciplinary representations, gesture analysis reveals their communication efforts by highlighting critical features within representations.

> When students struggle to map the features of one type of representation onto those of another and explain in words or text the relationship between them simultaneously, gesture analysis reveals students' efforts to re-represent disciplinary representations in series and consider salient or recognized features while ignoring other features.

Gestures capacity to represent mechanisms and reveal students' developing RC extends across STEM disciplines. Consider the following description of a students' gesture during classroom discourse in a physics laboratory:

> One measure of the gesture's significance is that her statement is unintelligible without it. Another measure of the gesture's significance is that it is an intuitively compelling expression of Jenny's thinking about the motion, and the [other] participants treat it as such; at Jenny's gesture, the TA stops talking, and within a few seconds the group reaches the correct conclusion about the velocity at the top of the trajectory. Finally, the gesture is eloquent; it's hard to imagine words that would complete Jenny's sentence with anything like the clarity and brevity that the gesture provides (Scherr 2008).

We agree. Indeed, we demonstrated previously that students in other STEM disciplines use gesture in ways similar to our above conjectures and the argument by Scherr. When organic chemistry students translate among multiple molecular diagrams, their gestures reveal information about their RC. For instance, when students extend two fingers to re-represent bonds between atoms, we observe these students translate a molecular diagram into a representational gesture that highlights critical features of the disciplinary representation (Lira et al. 2012). Likewise, when students move their hands through space to represent a spatial transformation of a molecule they use representational gestures to illustrate dynamic processes difficult to communicate in speech. Whereas our prior gesture analyses used the technique for other purposes, we set out here to use gesture analysis to assess students' developing RC.

Analyzing Gesture to Assess Students' Developing RC

In a pilot investigation we assessed students' developing RC by analyzing the gestures they produced while engaged in mechanistic reasoning about a complex phenomenon in cellular physiology. The results presented here come from a larger data corpus collected to better understand how students learn from multi-representational technologies. In this investigation, we used a sampling method to select portions of the interview protocol that elicited both mechanistic reasoning and students' use of representational gestures to support their efforts to communicate their mechanistic reasoning. Our aim was to demonstrate the utility of gesture for assessing the development of RC among STEM students and offer illustrative cases to motivate future work that leverages gesture analysis to investigate students' developing RC.

Population Participants included 10 undergraduate students whose ages ranged from 20–23 years. All students were currently enrolled in a biological sciences course titled "Homeostasis: The physiology of plants and animals." Nine students declared biology as their major and one student declared biochemistry. All students had taken a minimum of two semesters of chemistry and eight had a minimum of one semester of physics (mechanics) and calculus.

Investigation Context The course, Homeostasis, emphasized physiological mechanisms essential to the survival of multicellular organisms. Students were recruited from this course because of the emphasis the course placed on mechanistic reasoning and the high frequency of external representations presented in curricular materials.

For the investigation, we interviewed students about their understanding of a core concept, the *resting membrane potential*, taught in the homeostasis course. Biologists model the membrane potential as an electrical circuit. They therefore borrow representations from physics, chemistry, and mathematics. Each representation highlights a facet of the concept. For instance, diagrams illustrate the physical structure of the membrane and how it permits or prohibits molecular processes such as diffusion. Graphs of the membrane potential illustrate aggregated, quantitative trends in magnitude and direction of the voltage as a function of time. This disciplinary concept therefore lends itself well to investigating students' developing RC because mastering the concept requires students to use representations to think and communicate.

Cells generate resting membrane potentials via two compulsory initial conditions: concentration gradients and selectively permeable membranes. First, molecular motor proteins called sodium-potassium ATP-ases transfer bond energy from molecules of ATP to do the work of establishing concentration gradients for these ionic species. A concentration gradient for an ionic species creates a chemical driving force[1] (see Fig. 1). Second, ion-selective transmembrane proteins called passive-leak channels establish a selectively permeable membrane. The cell's permeability is often highest for potassium ions and thus these ions will diffuse from their area of greater concentration to their area of lesser concentration more so than all other ions. Large, negatively charged aqueous proteins are impermeable to the membrane. Negatively charged chloride ions distribute themselves passively. Active transport, thus, establishes concentration gradients whereas passive transport generates electrical gradients called membrane potentials measured as voltages. Membrane potentials refer to the separation of oppositely charged particles at the membrane. The membrane potential "rests" when all electrical and chemical driving forces are equal and opposite and the cell observes no next flux for all ionic species (i.e. dynamic equilibrium is reached).

---

[1] Entropy drives diffusion. Thus, the "chemical driving force" is not a true force. Physiologists use this metaphor because it supports intuition and communication. Also, mathematical and physical modelling support its use by construing the solvent as exerting a viscous force (drag) upon which work is done during diffusion (Weiss 1996).

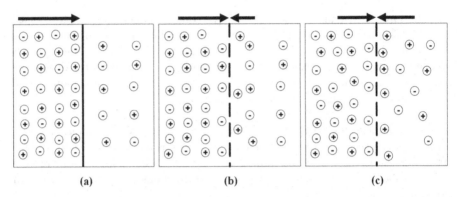

Fig. 1 A schematic diagram simplified to include only two solutions of potassium chloride (aqueous) to represent the generation of the resting membrane potential (**a**) The same numbers of oppositely charged potassium (+) and chloride (−) ions make the solutions inside and outside the cell electroneutral (i.e. 0 volts). The arrow indicates that the chemical driving force directs the potassium ions to the right. (**b**) A membrane selectively permeable (illustrated with white space) to positively charged potassium ions permits the ions to diffuse from areas of greater concentration to areas of lesser concentration via passive transport (i.e. diffusion). Note the electrical driving force in the direction opposite to the chemical driving force. (**c**) The system will reach dynamic equilibrium when the electrical driving force is equal but opposite the chemical driving force. Here the inside of the cell holds a negative voltage relative to the outside because of the excess negatively charged chloride ions in its intracellular solution; water not represented

Protocol The present investigation took place outside the course itself. Data was collected in a laboratory setting on the university campus. Students were recruited via email. Interviews were held during weeks 5 through 13 of a 16-week semester. The first investigator conducted all interviews. Interviews lasted no longer than 90 min.

The protocol consisted of four tasks: an explanation task, a drawing task, a learning task, and a second explanation task. In the first explanation, we asked students to explain how a cell generates a resting membrane potential. In the second task, we elicited student-generated diagrams from students. We chose this task to determine themes that recurred across students' first verbal explanations and diagrammatic explanations prior to any instruction that we delivered. In the third task, we delivered to students a learning experience with a multi-representational technology. We will not introduce the technology here or make claims regarding how students learn from this technology. We aim to assess students' developing RC not influence it but we have provided this information here to illustrate the general context of the data collection. In the fourth and final task, we asked students to again explain how a cell generates a resting membrane potential.

Analysis We conducted a constant comparative method (Glaser 1965) to contrast the information represented in student diagrams and explanations. Regarding the diagrams, we cataloged the diagrams representation of the entities (e.g. potassium ions), their properties (e.g. charge), and processes (e.g. diffusion), and interactions (e.g. attractions). We leveraged the framework of the discipline to begin this analysis. For instance, using disciplinary knowledge we documented when students represented passive or active transport mechanisms in their diagram. Next, we analyzed video by turning off students' speech and identifying all episodes when students appeared to produce representational gestures (i.e. when their gesture window expanded beyond the extent of their rhythmic beat gestures). We then analyzed their speech to determine the concepts that corresponded to their gestures. After all representational gestures were identified and categorized by disciplinary content, we reiterated the process until we isolated 15 episodes. Our taxonomy for classifying students' representational gestures during mechanistic reasoning provided three kinds of illustrative episodes of students representing information in gesture that was not detectable in their disciplinary diagrams.

Specific to gesture analysis, we followed the analytic process described by McNeill (1992, 2005). This process involved establishing when a gesture begins, when the full structure of the gesture peaks, and when it ends. The three phases are referred to as the **preparation**, the **mid-stroke**, and the **retraction** phases, respectively. We will utilize *italics* to indicate the word emphasized by the student during the co-occurring gesture at approximately mid-stroke.

Results

Describing Entities and Processes

Our first illustrative episode shows how students used gesture to represent unseen entities and processes critical to the generation of the resting membrane potential. Recall that many observable phenomena in science are understood in reductionist terms whereby unseen entities and processes interact to form emergent macroscopic or measurable properties. Because such entities and processes extend beyond human perception gesture constitutes an appropriate mode to represent this information because gesture unfolds in three-dimensional space over time.

In this first episode, one student, Allan, toward the end of his first explanation, begins to describe equilibrium in terms of sub-microscopic entities and processes. He first describes one kind of entity, potassium ions, in terms of their net charge. In nature, potassium often bonds with other elements to form salts. When placed in water, potassium dissociates from negatively charged elements and exists as a cation in solution. This is true in physiological solutions. Allan leverages his knowledge of this entity to reason through this explanation task (see Fig. 2).

Then, Allan describes a critical step in the process when he argues that because of potassium ions chemical properties (i.e. their chemical identity) they will "want to leave" (left panel) but because of their physical properties (i.e. their electrical charge) they will leave to a certain extent that keeps them in equilibrium (right

Fig. 2 Allan: So potassium, since it's plus [**preparation** stroke: at chest, left hand contracts fingers to form fist], it will want to *leave* [**mid-stroke**: thumb extends and points behind the body as illustrated in the left panel] but [**preparation** stroke: forefingers extend with palm facing the body] it *leaves* to the extent [**mid-stroke**: hand rocks to the left and right gradually decreasing in amplitude as illustrated in the right panel] where it keeps in equilibrium [**retraction**: hand pauses at chest and a new gesture emerges]. So the concentration gradient is pushing outward where[as] the electrochemical gradient is pushing it inwards

panel). We see the process of dynamic equilibrium represented by Allan when he gestures to represents motion both into and out of the cell.

Allan concludes his explanation by noting the importance of transmembrane proteins referred to as "channels" (see Fig. 3). Recall that channels refer to microscopic entities studded throughout the cell's membrane; they permit the diffusion of ions and thus the generation of the resting membrane potential. By referring to channels, Allan has invoked an entity as a referent in his speech. Allan proceeds to represent the entity with a gesture (see Fig. 4).

During the mid-stroke of the gesture, Allan represents a closed channel with a closed fist (left panel). Then, he contrasts the closed state of the channel by opening his fist into an "O" shape during the retraction phase of the gesture (right panel). Allan has thus described underlying entities and processes by representing them with gesture.

Allan next completed the drawing task. Notice that Allan constructed a single diagram (see Fig. 4) to represent the cell, as opposed to a series of diagrams that represent the process of the cell generating a resting membrane potential (see Fig. 1). By representing a process with one diagram, Allan runs into trouble. He fails to represent a clear mechanism for the generation of the *resting* membrane potential. He also fails to represent the process of dynamic equilibrium. The series ought to include three stages: an electroneutral cell with a concentration gradient for potassium ions, the mechanism of passive transport through potassium leak chan-

Fig. 3 Allan: [...] the channels are going with the concentration gradient and the electrochemical gradient [**preparation**: right hand raises from waist to chest and fist is closing] and *keeping* [**mid-stroke**: closed fist] those [**retraction**: fist is opening as hand falls toward floor] closed or open

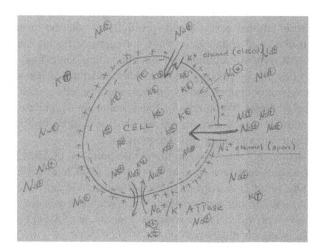

Fig. 4 Allan's diagram. A single frame that represents multiple transport mechanisms and non-equilibrium conditions

nels, and the separation of oppositely charged ions across the membrane. Allan represents this last stage with plus and minus signs located near the membrane but he fails to represent the first two stages.

Next, notice that at the top of his diagram Allan uses arrows to represent motion through what he labels as a potassium (K^+) channel (see Fig. 4). We presume that Allan means for these arrows to represent vectors because they vary in direction and magnitude but because he does not label them we cannot be certain. What is certain is that his diagram does not indicate dynamic equilibrium or a "resting" membrane potential because the arrows differ in size and this suggests a net flux of ions. We see the same problem at the middle, right side of diagram where we observe one large vector directed inward and no vector in the opposite direction. These vectors sum to a net inward flux of positive ions and thus fail to represent dynamic equilibrium.

Last, notice that Allan uses text to note the open and closed state of channels but he fails to represent the closed channels that he represented in gesture. Whereas Allan coordinated his speech and gesture to add specificity to the referent of a closed channel, Allan fails to coordinate his writing and drawing to represent a closed channel.

Allan's diagram represents the underlying entities, processes, and mechanisms responsible for the resting membrane potential but his diagram underspecifies the mechanism and the sequence of events. First, Allan did not clearly represent the transport mechanism. He included both passive and active transport mechanisms in his diagram. We can only assume that both play an equal role in the generation of the resting membrane potential. They don't. Second, Allan does not represent the temporal sequence of the mechanism correctly in his diagram. Allan represents the final conditions by separating positive and negative charges in his diagram but he does not represent prior stages that allow us to distinguish between cause and effect.

Allan's representational gestures add specificity to what he means by equilibrium and open or closed channels in ways not captured by his diagram. Allan's gestures represent dynamic equilibrium by illustrating ionic flux equal in both directions; he does this by rocking his hand back and forth. Allan's gestures also represent channels in both open and closed states and he tells us that channels "are going" with the gradients. We interpret Allan's statement as a reference to passive transport and thus in speech and gesture we observe a more favorable picture of Allan's RC. His gestures promote this interpretation.

We do not wish to adopt a naïve and literal interpretation of Allan's diagram. We are aware that images represent time in many ways (McCloud 2006; Schnotz and Lowe 2008). Our point is that information not detected in our assessment of his diagram was detected in our assessment of his speech and gesture. We argue that this gesture analysis provided a more sensitive assessment of Allan's *developing* RC compared to assessing his diagram alone.

Identifying Critical Features in Representations

In this next episode, we illustrate how a students' gesture identifies a critical feature in an external representation that supports explanation. Here, we observe Carrie, who fails to specify the accurate transport mechanism in her diagram (see Fig. 5). Carrie, like Allan, represents both active (represented at bottom center) and passive transport (represented at top right) in a single-frame diagram. If anything, Carrie's

Fig. 5 Carrie's diagram. A single frame that represents multiple transport mechanisms and non-equilibrium conditions

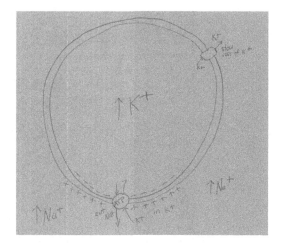

diagram indicates that active transport generates the membrane potential because she represents the separation of opposite charges at the bottom of the diagram near the active transport site but nowhere else.

Regarding the "slow loss of K^+", Carrie represents passive transport as an effect with no clear cause. We might be generous and infer that the arrow next to the K^+ in the center of her diagram means that a more concentrated solution of potassium ions resides in the intracellular fluid while a less concentrated solution of potassium ions resides in the extracellular fluid and that these initial conditions lead to the outward diffusion of potassium ions. We agree that Carrie's diagram likely represents this information—albeit with one frame as opposed to three. This interpretation, however, means that Carrie has used the same symbol, an arrow, to convey three distinct concepts—extent or magnitude, flow or passive transport, and a chemical reaction or active transport while providing text once to support communication. Although Carrie's skill at linking the arrow symbol to multiple concepts hints at her developing RC, her failure to coordinate text with her manifold use of the arrows leads us to adopt a less favorable view.

In the second explanation task, Carrie resolves the discrepancy. During the mid-stroke of Carrie's gesture, we see her place her hands in opposition to one another (see Fig. 6). At this point she remarks, "Those two forces kind of reach an equilibrium." With her gesture, Carrie represents an interaction between abstract entities—Carrie's speech and gesture describe and represent a cause (i.e. forces) for the effect (i.e. changes in ionic flux). Her hands waver back and forth until they come to a stop during the retraction phase of the gesture. Thus, Carrie identifies the critical features of these entities—their direction and magnitude—and selects an appropriate mode to illustrate how dynamic equilibrium occurs.

In Carrie's diagram, she used arrows to represent the diffusion of potassium ions but she failed to specify a cause. Carrie's gesture adds specificity to her explanation in several ways. First, her equal but opposite representational gesture describes the

Fig. 6 Carrie: [...] there's less potassium outside so it wants to travel out but at the same time that creates a more negative internal environment so it wants to come back in. [**preparation**: hands move toward mid-line to press against each other near the chest] So *between* [**mid-stroke**: hands wave near and far from the body] those two forces it kind of reaches an equilibrium [**retraction**: hands clasp near the chest] [...] and it maintains at −60 millivolts [...]

two forces and defines what "resting" means. The gesture also represents the physical quantity, −60 millivolts, Carrie mentioned in speech. Carrie represented a negative voltage in her diagram of the cell but we did not know from her diagram that we could characterize the separation of charge as a measured physical quantity (i.e. a voltage). Finally, Carrie's drawing represented the "slow loss" of potassium ions (K^+). It did not specify the bi-directional motion of potassium ions through a channel (i.e. dynamic equilibrium). While gesturing, Carrie explains that potassium ions "want" to both travel out and come back in. In the prior episode, Allan represented the effect of equal but opposite forces when his gesture represented ions moving equally in opposite directions. Carrie's gesture represents the cause.

Relating External Representations

This last episode provides insight into how students relate multiple external representations in gesture—students do so by re-representing disciplinary representations in tandem. When people talk, their co-occurring representational gestures transition seamlessly from one form to the next. As the visual component of thought, these gestures represent in series multiple scenes or multiple features of one scene

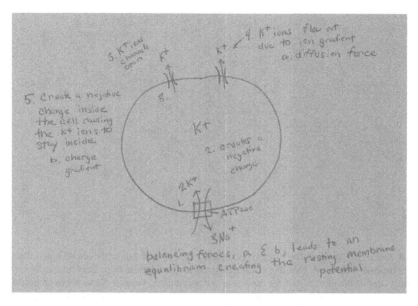

Fig. 7 Oscar's diagram. A single-frame sequenced by numbered stages with supporting text that describes the mechanism responsible for the final dynamic equilibrium conditions

(McNeill 1992). When students represent science concepts with gesture they are therefore afforded the opportunity to describe processes, interactions, and features of entities in sequence as they occur in thought.

In this episode, one student, Oscar, gestures to re-represent and relate a mechanistic interaction to a voltage-time graph. Unlike Carrie, Oscar mentioned opposing forces in his first explanation and in his diagram but forces were represented with text as opposed to vector notation (see Figs. 1 & 4). The arrows in his diagram were labeled motion not forces. Moreover, he did not relate his diagram to a graph (Fig. 7).

In Oscar's second explanation, we learn more about his RC. First, notice that Oscar, like Carrie, produces a gesture that represents an interaction between two forces (see Fig. 8). In speech, Oscar explains that the membrane potential is resting because the forces are balancing at a precise voltage. During the mid-stroke of the first gesture, he depicts how the forces balance by waving his hands left to right in front of his body until his hands stop during the retraction phase (left panel). With this gesture, Oscar represents the specific and critical mechanistic interaction responsible for a membrane potential "resting" and selects an appropriate mode to represent how dynamic equilibrium occurs. Oscar's gestures add specificity to his verbal explanation by representing the direction and magnitude of the forces much in the same way that Carrie's episode illustrated. He represents equilibrium with equal and opposite vectors to the extent that gestures can represent such abstract entities.

What is more interesting about Oscar's episode, in contrast to Carrie's, is that when he mentions the "equilibrium" of potassium ions, he elaborates by relating his

Fig. 8 Oscar: So [**preparation**: right hand meets left hand mid-line at the chest] *balancing* those forces [**mid-stroke**: hands rock to the left and right] out [**retraction**: right hand relaxes as left hand begins to depict the direction of the motion of potassium ions] because K⁺ wants to go outside the cell because there's a high concentration within the cell and balancing those forces again, like I answered in the beginning, creates the resting membrane potential which is closer to the [**preparation**: hands lift to meet mid-line just below the chest] K^+ ion [**mid-stroke**: hands separate away from each other in a line parallel to the floor] equilibrium [**retraction**: hands repose to a clasp at the chest]

first gesture that represented equilibrium to a second gesture that represents a constant voltage (right panel). Oscar remarks that the resting membrane potential will be closer to the equilibrium (potential) of potassium ions. During the mid-stroke of the second gesture, Oscar represents a critical feature of the graph that he experienced during the learning task that occurred before the second explanation task. He represents the critical feature by flattening both of his palms parallel to the floor and sweeping his hands away from his body to trace a straight line with a slope of 0. By creating two distinct representational gestures in series, Oscar demonstrates his developing RC when he re-represents two critical components of external representations in gesture and relates them in speech (Fig. 8).

In contrast to his diagram, Oscar's speech and gesture provide a more favorable view of his understanding of how to represent dynamic equilibrium and the resting membrane potential. Oscar notes "equilibrium" in the text that accompanies his diagram but he does not represent dynamic equilibrium satisfactorily because he represents only the flow of potassium ions out of the cell. He does not represent a net flux of zero. In speech and gesture, however, Oscar represents two opposing forces to define "resting" as dynamic equilibrium.

Conclusions

In this chapter we presented a framework for using gesture to analyze students' developing RC in STEM disciplines. We illustrated the application of this framework with three episodes that highlight students' developing RC while they explained a cellular physiology concept. The results lend support to efforts that seek to further understand the role of gesture in assessing students' developing RC. We highlighted that students' gestures have the capacity to reveal students' RC in ways not captured by student-generated diagrams or speech alone.

Our multi-modal assessment provided a window to see how students' co-speech gestures offer a picture of their developing RC that differs from the formal disciplinary representations they construct. Students' diagrams did not represent time clearly. Because students constructed single frame diagrams for multi-stage mechanisms, their representations appeared as smatterings that obscured sequence and causality. Constructing multiple diagrams to highlight sequenced events aids in communicating meaningful relations (Agrawala et al. 2003). Even when provided with a series of written sentences, biology students struggle to organize the sequence in the correct order (Zohar and Tamir 1991). Analyzing students' co-speech gestures, however, shows that students describe and represent dynamic equilibrium near the end of their explanations. Thus, students' reveal their developing RC by coordinating their speech with gestures that represent spatio-temporal information.

Single frame diagrams can represent multi-stage processes but to communicate this information well the diagrams need supporting text. We found that when students failed to accompany their diagrams with supporting text, they failed to communicate the interactions between entities that function as critical steps in mechanisms. Prior approaches demonstrated that non-biology students vary the amount of text they provide when sketching physiology diagrams (Ainsworth et al. 2007). Sketching to explain to another person encourages students to use text in their diagrams. Students in this investigation were told to construct a self-explanatory drawing that a high school student could understand. Although not the final word, it appears that as physiology students learn to construct representations of mechanisms they sequence events first with text not diagrams. Students' gestures, however, highlighted interactions between forces and illustrated how these interactions were responsible for processes such as ionic flux or dynamic equilibrium. By identifying the features that students omit when they sketch diagrams and the features they represent with speech and gesture, we have extended our knowledge of students' developing RC and how gesture analysis complements other assessment efforts.

The present investigation does not reveal the representativeness of these episodes. Although we were able to identify these episodes in a small data set (n = 10) we acknowledge that the students are in upper division biology and may differ from, say, younger students learning in different STEM disciplines. Future investigations may overcome this limitation by adopting a meta-representational approach (diSessa 2004). For instance, investigators could select a cross-section of STEM students rather than students in one discipline. Asking students to first complete a standardized drawing task and then complete an explanation task would provide a more rigorous test to determine if gesture provides information absent in students' diagrams.

Second, our gesture analysis revealed students' developing RC but the design prevents us from understanding specifically how gesture production develops in concert with student-generated representations. In this study, we assessed diagram construction at one time point but verbal explanation at two time points. Future investigations could over come this limitation by executing a counter-balanced design that assesses students' gesture production and diagram construction an equal number of times.

Our multi-modal assessment leveraged the affordances of diagram construction and explanation tasks to provide a window into students' developing RC. By analyzing students' gestures we produced a more complete picture of how students' RC develops in relation to that of one formal disciplinary representation—the diagram. We acknowledge that much remains unknown regarding how these two modes of communication develop in step. Nevertheless, these results provide a proof of concept that students' gestures highlight different disciplinary concepts and representational features than their diagrams and thus support efforts to assess students' developing RC with gesture analysis.

References

Agrawala, M., Phan, D., Heiser, J., Haymaker, J., Klingner, J., Hanrahan, P., & Tversky, B. (2003). Designing effective step-by-step assembly instructions. *ACM Transactions on Graphics (TOG), 22*(3), 828–837.

Ainsworth, S. (2006). DeFT: A conceptual framework for considering learning with multiple representations. *Learning and Instruction, 16*(3), 183–198.

Ainsworth, S., Galpin, J., & Musgrove, S. (2007). Learning about dynamic systems by drawing for yourself and for others. *In EARLI conference 2007*.

Alibali, M. W. (2005). Gesture in spatial cognition: Expressing, communicating, and thinking about spatial information. *Spatial Cognition and Computation, 5*(4), 307–331.

Alibali, M. W., & Nathan, M. J. (2012). Embodiment in mathematics teaching and learning evidence from learners' and teachers' gestures. *Journal of the Learning Sciences, 21*(2), 247–286.

Azevedo, F. S. (2000). Designing representations of terrain: A study in meta-representational competence. *The Journal of Mathematical Behavior, 19*(4), 443–480.

Becvar, L. A., Hollan, J., & Hutchins, E. (2005). Hands as molecules: Representational gestures used for developing theory in a scientific laboratory. *Semiotica, 156*, 89–112.

Chi, M. T., Feltovich, P. J., & Glaser, R. (1981). Categorization and representation of physics problems by experts and novices*. *Cognitive Science, 5*(2), 121–152.

Chu, M., & Kita, S. (2011). The nature of gestures' beneficial role in spatial problem solving. *Journal of Experimental Psychology: General, 140*(1), 102.

Church, R. B., & Goldin-Meadow, S. (1986). The mismatch between gesture and speech as an index of transitional knowledge. *Cognition, 23*(1), 43–71.

Cook, M., Carter, G., & Wiebe, E. N. (2008). The interpretation of cellular transport graphics by students with low and high prior knowledge. *International Journal of Science Education, 30*(2), 239–261.

Crick, F. H., & Watson, J. D. (1954). The complementary structure of deoxyribonucleic acid. *Proceedings of the Royal Society of London Series A: Mathematical and Physical Sciences, 223*(1152), 80–96.

Demir, Ö. E. (2009). *A tale of two hands: Development of narrative structure in children's speech and gesture and its relation to later reading skill* (Doctoral dissertation). Retrieved from ProQuest. (Accession No. 3369323).

Ehrlich, S. B., Levine, S. C., & Goldin-Meadow, S. (2006). The importance of gesture in children's spatial reasoning. *Developmental Psychology, 42*(6), 1259.

Flood, V. J., Amar, F. G., Nemirovsky, R., Harrer, B. W., Bruce, M. R., & Wittmann, M. C. (2014). Paying attention to gesture when students talk chemistry: Interactional resources for responsive teaching. *Journal of Chemical Education, 92*(1), 11–22.

Garber, P., & Goldin-Meadow, S. (2002). Gesture offers insight into problem-solving in adults and children. *Cognitive Science, 26*(6), 817–831.

Glaser, B. G. (1965). The constant comparative method of qualitative analysis. *Social.problems, 12*(4), 436–445.

Göksun, T., Hirsh-Pasek, K., & Golinkoff, R. M. (2010). How do preschoolers express cause in gesture and speech? *Cognitive Development, 25*(1), 56–68.

Hammer, D., Sherin, B., & Kolpakowski, T. (1991). Inventing graphing: Meta-representational expertise in children. *Journal of Mathematical Behavior, 10*(2), 117–160.

Hinze, S. R., Rapp, D. N., Williamson, V. M., Shultz, M. J., Deslongchamps, G., & Williamson, K. C. (2013). Beyond ball-and-stick: Students' processing of novel STEM visualizations. *Learning and Instruction, 26*, 12–21.

Iverson, J. M., & Goldin-Meadow, S. (2005). Gesture paves the way for language development. *Psychological Science, 16*(5), 367–371.

Johnstone, A. H. (1991). Why is science difficult to learn? Things are seldom what they seem. *Journal of Computer Assisted Learning, 7*(2), 75–83.

Jolley, R. P. (2010). *Children and pictures: Drawing and understanding*. Chichester: Wiley.

Kastens, K. A., Agrawal, S., & Liben, L. S. (2008). Research methodologies in science.education: The role of gestures in geoscience teaching and learning. *Journal of Geoscience Education, 56*(4), 362–368.

Kozma, R. B., & Russell, J. (1997). Multimedia and understanding: Expert and novice responses to different representations of chemical phenomena. *Journal of Research in Science Teaching, 34*(9), 949–968.

Kozma, R., & Russell, J. (2005). Students becoming chemists: Developing representational competence. In J. K. Gilbert (Ed.), *Visualization in science education* (pp. 121–145).

Latour, B. (1986). Visualization and cognition. *Knowledge and Society, 6*, 1–40.

Lira, M., Stieff, M., & Scopelitis, S. (2012). The role of gesture in solving spatial problems in STEM. In J. van Aalst, K. Thompson, M. J. Jacobson, & P. Reimann (Eds.), *The future of learning: Proceedings of the tenth international conference of the learning sciences (ICLS) - volume 2, short papers, Symposia, and abstracts* (pp. 406–410). Sydney: International Society of the Learning Sciences.

McCloud, S. (2006). *Making comics: Storytelling secrets of comics, manga, and graphic novels*. William Morrow.

McDermott, L. C., Rosenquist, M. L., & Van Zee, E. H. (1987). Student difficulties in connecting graphs and physics: Examples from kinematics. *American Journal of Physics, 55*(6), 503–513.

McNeill, D. (1985). So you think gestures are nonverbal? *Psychological Review, 92*(3), 350.

McNeill, D. (1992). *Hand and mind: What gestures reveal about thought*. University of Chicago Press.

McNeill, D. (2005). *Gesture and thought*. University of Chicago Press.

Nitz, S., Ainsworth, S. E., Nerdel, C., & Prechtl, H. (2014). Do student perceptions of teaching predict the development of representational competence and biological knowledge? *Learning and Instruction, 31*, 13–22.

Parnafes, O. (2007). What does "fast" mean? Understanding the physical world through computational representations. *The Journal of the Learning Sciences, 16*(3), 415–450.

Pedelty, L. (1985). *Gestures in aphasia* (doctoral dissertation). Unpublished doctoral dissertation.

Perry, M., Church, R. B., & Goldin-Meadow, S. (1988). Transitional knowledge in the acquisition of concepts. *Cognitive Development, 3*(4), 359–400.

Roth, W. (2000). From gesture to scientific language. *Journal of Pragmatics, 32*(11), 1683–1714.

Scherr, R. E. (2008). Gesture analysis for physics education researchers. *Physical Review Special Topics-Physics Education Research, 4*(1). 010101.

Schnotz, W., & Lowe, R. (2008). A unified view of learning from animated and static graphics. In R. Lowe & W. Schnotz (Eds.), *Learning with animation research implications for design* (pp. 304–356). Cambridge: Cambridge University Press.

Schwartz, D. L., & Black, J. B. (1996). Shuttling between depictive models and abstract rules: Induction and fallback. *Cognitive Science, 20*(4), 457–497.

diSessa, A. A. (2004). Metarepresentation: Native competence and targets for instruction. *Cognition and Instruction, 22*(3), 293–331.

Shah, P., & Hoeffner, J. (2002). Review of graph comprehension research: Implications for instruction. *Educational Psychology Review, 14*(1), 47–69.

Sherin, B. L. (2000). How students invent representations of motion: A genetic account. *The Journal of Mathematical Behavior, 19*(4), 399–441.

Srivastava, A., & Ramadas, J. (2013). Analogy and gesture for mental visualization of DNA structure. In D. F. Treagust & C. Y. Tsui (Eds.), *Multiple representations in biological education* (pp. 311–329). New York: Springer.

Stieff, M. (2011). When is a molecule three dimensional? A task-specific role for imagisticreasoning in advanced chemistry. *Science Education, 95*(2), 310–336.

Stieff, M., Hegarty, M., & Deslongchamps, G. (2011). Identifying representational competence with multi-representational displays. *Cognition and Instruction, 29*(1), 123–145.

Uttal, D. H., & O'Doherty, K. (2008). Comprehending and learning from 'visualizations': A developmental perspective. In J. K. Gilbert, M. Reiner, & M. Nakhleh (Eds.), *Visualization: Theory and practice in science education* (pp. 53–72). Dordrecht: Springer.

Weiss, T. F. (1996). *Cellular biophysics* (Vol. 1). Cambridge: MIT press.

Zohar, A., & Tamir, P. (1991). Assessing students' difficulties in causal reasoning in biology—a diagnostic instrument. *Journal of Biological Education, 25*(4), 302–307.

Assessing Representational Competence with Eye Tracking Technology

Inga Ubben, Sandra Nitz, Kristy L. Daniel, and Annette Upmeier zu Belzen

Introduction

Science classes breathe from the use of models in the form of various representations. Representational competence (RC), the ability to think about, use, and to reflect underlying processes and characteristics of representations (Kozma et al. 2000; Kozma and Russell 1997; Kozma and Russell 2005) is thus an elementary and crucial skill for everyone involved – students as well as teachers. RC is highly context-specific which means that an individual can be at different stages dependent on the representation's content (Kozma and Russell 2005). From a scientific point of view this means that the state of RC and its development cannot be generalized but need to be investigated separately for different contents. This has been done elaborately with interviews, questionnaires, or other methods (e.g., Nitz et al. 2014; Stieff 2007; Tippett and Yore 2011). The highly visual character of external representations yet led to an increasing number of studies using eye tracking technology to literally make visual perception of representations visible. This reaches from expert-novice comparisons (e.g., Jarodzka et al. 2010) to the use of multi-representational displays (Stieff et al. 2011) to the influence of representational form on RC (e.g., Novick et al. 2012) just to name a few examples. The attraction of this method is that it "provides objective and quantitative evidence of the user's

I. Ubben (✉) · A. Upmeier zu Belzen
Humboldt-Universität zu Berlin, Berlin, Germany
e-mail: inga.ubben@biologie.hu-berlin.de; annette.upmeier@biologie.hu-berlin.de

S. Nitz
University of Koblenz and Landau, Landau, Germany
e-mail: nitz@uni-landau.de

K. L. Daniel
Texas State University, San Marcos, TX, USA
e-mail: kristydaniel@txstate.edu

© Springer International Publishing AG, part of Springer Nature 2018
K. L. Daniel (ed.), *Towards a Framework for Representational Competence in Science Education*, Models and Modeling in Science Education 11,
https://doi.org/10.1007/978-3-319-89945-9_11

visual and (overt) attentional processes" (Duchowski 2002, p. 455). This chapter will describe how eye tracking technology combined with verbal methods offers us the opportunity to get a more comprehensive insight into cognitive processes of RC using the example of phylogenetic trees as biology-specific representations.

Representational Competence and Model Competence

Kozma and Russell (2005) defined different skills and five stages or levels of RC for chemistry reaching from basic RC to expert use:[1] Individuals on the first level (representation as depiction) generate representations as depiction of a phenomenon's physical features. On the second level (symbolic use), individuals read representations superficially without paying attention to the representation's syntax or semantics. The third level (syntactical use) is characterized by reading representations on a syntax level without considering the underlying meaning. Furthermore, two representations of the same phenomenon are compared only by syntax and superficial features. On the fourth level (semantic use) individuals connect surface features of representations to the represented content and consider the meaning of the representation. Individuals on this level compare different representations based on the represented meaning and are, thus, able to recognize the shared underlying meaning of several different representations of the same phenomenon and to transfer information between these representations. Moreover, this level is characterized by spontaneous use of representations for explanation. Individuals on level five (reflective, rhetorical use) can explain physical phenomena with the help of one or several representations, support claims referring to certain features of a representation, select the most suitable representation, and explain their choice (Kozma and Russell 2005).

As we can see, reading representations gets more and more elaborated over levels: Low levels are characterized by using only superficial features of representations whereas higher levels include the underlying meaning (cf. Chi et al. 1981). Individuals on higher levels are able to deal with several representations and are able to reason about which is the most appropriate one. It is important to consider that different representational forms do not illustrate different ideas but are various representations of one underlying model (Passmore et al. 2014). According to Nersessian (2002) these representational forms of a model foster reasoning about the model by highlighting the salient characteristics. Furthermore, representations of a model are fruitful for communication between individuals and thus can lead to new insights (Nersessian 2002). A model *of* a phenomenon can be externalized in several representational forms which then can serve to use the model *for* predictions

---

[1] The five levels originally also include information about building representations. To focus on the visual aspect, this section will only mention aspects of reading representations besides level 1 for which no information about reading is provided. For a complete description of levels see Kozma and Russell (2005), p. 133.

and reasoning (Mahr 2008, 2009; Passmore et al. 2014). In the first case models are used medially, in the second methodically (Gilbert 1991; Mahr 2009). Handling models can take place on three different levels as described by the framework of model competence (Grünkorn et al. 2013; Upmeier zu Belzen and Krüger 2010): Level I and II cover the medial aspect of models while the more elaborated level III includes the methodical use.

Basics of Eye Tracking Technology

Eye tracking technology is widely used in several disciplines such as marketing research, psychology, and medicine. In the past years, it has also been more and more applied in science education research. It found its beginning in the nineteenth century with complicated mechanical apparatuses and was permanently improved over decades. Eye tracking devices included inter alia contact lenses and were thus quite invasive.[2] Modern eye trackers are far less invasive since they use infrared cameras to monitor eye movements. Several different techniques are available ranging from head-mounted systems over remote systems to lightweight mobile glasses.[3] Data are most often recorded by computers but also data collection using smartphones is available. Although the number of different techniques and manufacturers is continuously growing, the underlying principles remain mostly the same. The following section will give an overview of the mechanisms of eye movements, visual attention, measures used for research, and complementary data recording.[4]

Eye-Mind Assumption, Eye Movements, and Measures

When we visually perceive a situation, a representation, or a text, several cognitive processes filter relevant from irrelevant visual information. The combination of those processes is called visual attention. Visual attention can either occur overtly (when directly looking at something) or covertly (paying attention to something without moving the eyes to it, e.g., parafoveal attention; McMains and Kastner 2009). Eye tracking is based on the assumption that eye fixation on a locus indicates that visual attention lies on this locus and that thereby this locus is processed (eye-mind assumption; Just and Carpenter 1980). Since this definition excludes covert attention, Rayner (1998) assumes that visual attention and eye fixation might be

---

[2] For a detailed history of eye tracking see e.g., Duchowski 2007, pp. 51–59

[3] This chapter will mainly focus on remote systems.

[4] For an extensive introduction into theory and practice of eye tracking technology see Duchowski (2007) and Holmqvist et al. (2011).

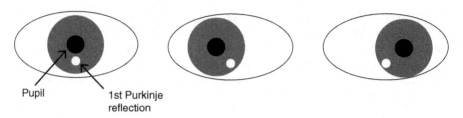

Pupil   1st Purkinje
                         reflection

Fig. 1 The vector between pupil and first Purkinje reflection changes with eye movement. Adapted from Duchowski 2007, p. 58, with permission of Springer

tightly linked at least in complex tasks when information need to be processed like for example during reading.[5]

Fixations are not static but miniature movements of the eye that stabilize the retina over a locus of interest. Through this the locus can be visually perceived. Saccades, in contrast, are fast eye movements between two fixations with the aim of repositioning the retina onto another locus of interest. During saccades, there is no visual perception (Duchowski 2007).[6]

How can we actually measure those eye movements described above? Video-based eye trackers capture eye movements based on pupil position and corneal reflection of infrared light. Especially in remote systems, these two parameters are necessary to compensate for head movements relative to the static stimulus monitor. More precisely, the light of an infrared source underneath the stimulus monitor is reflected by the eye and the reflection is imaged by a camera. While the relative position of the pupil to the stimulus screen and hence to the infrared source changes due to eye movements, the relative position of the corneal reflection (also called first Purkinje reflection) on the surface of the eye ball remains the same. Hence, the vector between pupil and corneal reflection can be used to calculate eye movement in relation to the stimulus screen and the infrared source, respectively (see Fig. 1; Duchowski 2007).

---

[5] Duchowski (2007) points out that the eye-mind assumption might be limited and - under ideal conditions – eye tracking should be complemented by brain activity measures. He gives the example of astronomers who purposefully separate attention from gaze direction when they search for faint stars which cannot be spotted when directly looking at them. Furthermore, experts are able to perceive details of representations parafoveally as described in "Assessing Differences between Experts and Novices" of this chapter.

[6] There are also other types of eye movements like smooth pursuits (the eye follows a moving object) and nystagmus (counterbalancing head movements) which do not play a role in assessing RC.

How to Interpret and Use Eye Tracking Data

Fixations are an important measure in eye tracking technology since they indicate – according to the eye-mind-assumption (Just and Carpenter 1980) – to what feature of a representation an individual points attention to. To make sense of fixations, areas of interest (AOIs) are created corresponding to the research question. These could be relevant or irrelevant parts of a representation to solve a certain task. Subsequently, the number or duration of fixations in these AOIs can be compared. For example in a study about perception of fish locomotion, the fish's body was divided into different AOIs (fins, body, eyes etc.). These AOIs were inter alia relevant to describe the locomotion or not and number and duration of fixations were taken as indicators for expertise (Jarodzka et al. 2010). Besides that, time until first fixation of an AOI can give insights into how fast individuals can identify relevant features of a representation.

The sum of saccades, interrupted by fixations, results in the scanpath, the "route of oculomotor events through space within a certain timespan" (Holmqvist et al. 2011, p. 254). Hence, scanpaths represent patterns of eye movements and can be used inter alia for a preliminary data overview and cued retrospective think aloud which will be described later in this chapter. For scanpath comparison between subjects or groups, several approaches can be used as specified by Holmqvist et al. (2011).

Complementary Data

Eye movement data tell us, where and when participants fixated a locus but not why. The underlying cognitive processes can only be examined by triangulation with other methods such as verbal data. For this purpose, two different types of think aloud are recommended: Concurrent think aloud (CTA; Ericsson and Simon 1993; van Someren et al. 1994) and cued retrospective think aloud (cued RTA; van Gog et al. 2005). In a study on fault diagnosis CTA gave more valid insights into cognitive processes than RTA (Brinkman 1993). To overcome this difference in quality van Gog et al. (2005) expanded the method of RTA (Ericsson and Simon 1993) by cueing the RTA process via superimposition of the participant's own scanpath. In computer-based problem-solving tasks this cued RTA was found to be as valid as CTA by means of information about which actions were performed to solve a problem, why and how these actions were done, and how the participant reflected about performing them (cf. van Gog et al. 2005, p. 237). Furthermore, it offers the advantage that experts use a similar amount of words as novices when explaining their performance in contrast to CTA where experts verbalize less than novices (Jarodzka et al. 2010). Van Gog et al. (2005) recommend cued RTA to unite the advantages of CTA and RTA but call attention to issues to be investigated like individuals verbalizing different aspects of problem-solving in CTA and cued RTA and the influence of the tested group.

Stieff et al. (2011) addressed the important question whether verbal data and eye tracking data display the same cognitive processes, or more precisely if people mention the features of a multiple-representational display they look at or not. They could show quantitatively that – at least for CTA – there is a correlation between visual and verbal data so that for example missing data from eye tracking can be balanced by verbal data and vice versa. Furthermore - given the fact that both data sources display the same cognitive processes - they complement each other in terms of revealing different aspects of these processes. Eye tracking data for example show where exactly individuals look at whereas verbal data give insights into how they use a representation (cf. Stieff et al. 2011, p. 141). The question whether to use CTA or cued RTA needs to be decided case-dependently. In the study of Stieff et al. (2011) a head-mounted eye tracking device was used so that participants' head movements during CTA are less detrimental to data collection than with a remote system. For remote devices cued RTA might be more suitable since participants do not move caused by speaking during task completion or as a reaction to prompts by the experimenter.

Assessing Differences between Experts and Novices

Expertise, the "consistently superior performance on a specified set of representative tasks for a domain" (Ericsson and Lehmann 1996, p. 277), has been in the focus of numerous eye tracking studies (reviewed by Gegenfurtner et al. 2011). As described in the following, visual perception depends strongly on the degree of expertise.

Gegenfurtner et al. (2011) found in a meta-analysis of 65 eye tracking studies on expertise that experts have shorter fixation durations than novices because they need less time to put information into long-term memory and to retrieve it when needed (theory of long-term working memory; Ericsson and Kintsch 1995). Corresponding to the information-reduction hypothesis (Haider and Frensch 1999) experts fixate relevant AOIs more often and longer than irrelevant AOIs because they can distinguish faster between relevant and redundant AOIs and actively focus on relevant AOIs (Gegenfurtner et al. 2011). Furthermore, Gegenfurtner et al. (2011) could confirm the holistic model of image perception (Kundel et al. 2007) saying that experts do not need to directly focus on relevant features of a representation because of their higher visual span and parafoveal perception. This was expressed by shorter time to first fixation of relevant AOIs and longer saccades. The three above-named theories could be operationalized using different eye tracking measures such as fixation duration, fixation number, and time to first fixation on relevant and redundant AOIs as well as duration of saccades.

Regarding expert solving strategies in description of fish locomotion patterns, Jarodzka et al. (2010) found that there is not one expert nostrum. In fact, experts come to the same result with heterogeneous strategies. These strategies were operated by comparison of scanpaths (Levenshtein distance; Feusner and Lukoff 2008; for more details see Jarodzka et al. 2010). In the case of novices no strategies for describing fish locomotion could be found. This leads to the idea of using expert strategies to teach novices (eye movement modeling example, EMME; e.g., van Gog et al. 2009; Jarodzka et al. 2013; see "Future Application – How to Assess Representational Competence with Phylogenetic Trees Using Eye Tracking Technology"). Because of the heterogeneous expert strategies Jarodzka et al. (2010) propose to use the strategy of one certain expert instead of merging several strategies.

As Gegenfurtner et al. (2011) could show in their meta-analysis, experts are able to perceive information from representations without directly looking at them. This parafoveal or covert attention (McMains and Kastner 2009; see "Eye-Mind Assumption, Eye Movements, and Measures") has therefore to be kept in mind when analyzing eye tracking data. One possibility to uncover parafoveal attention might thus be its indirect detection via shorter time to first fixation of relevant AOIs and longer saccades as operated by Gegenfurtner et al. (2011). Furthermore, verbal data can give hints whether experts perceived features of representations without fixating them. In the case of RC it also has to be considered that the difference between experts and novices is stronger for schematic than for realistic stimuli (cf. Gegenfurtner et al. 2011).

Eye Tracking in Science Education

Eye tracking technology has been widely used for reading studies for more than 40 years (see Rayner et al. (2012) for a detailed overview) and there are several studies applying eye tracking for examining learning in the broadest sense (reviewed by Lai et al. 2013). A smaller amount of studies focused on RC with multi-media including complex graphical representations as described in the following. It was found that prior knowledge and providing new content knowledge respectively has beneficial influence on finding relevant information and interpreting the deeper meaning of a representation (Canham and Hegarty 2010; Cook et al. 2008). Furthermore visual cues in visualizations can guide the reader to relevant parts of a representation (Boucheix and Lowe 2010; Koning et al. 2010) and thereby can help with complex representations when participants are on a lower level of RC. In all four studies numbers of fixations on relevant representations and parts of representations, respectively, were taken as measures for visual perception and were supplemented with verbal data.

Especially in science education, students are confronted with various representational forms of invisible phenomena or structures such as diagrams, formulas, schemas, and many more. These are often highly abstract and scientific and therefore challenge novices (cf. Stieff et al. 2011, p. 123 f). Stieff et al. (2011) investigated the visual perception of multiple representations in chemistry education combining eye tracking with CTA. They showed four animated representational forms of one phenomenon at a time to college chemistry students. Participants got tasks which required choosing the most appropriate representational form(s) and to make claims and predictions with the help of these representational form(s). Hence, tasks necessitated a certain level of RC. It was found that students most often fixated the relevant representational form(s) and were also able to verbalize the correct answer to the task which lead to the conclusion that they are on an intermediate level of RC with chemistry specific animated representations as long as these are not too abstract (Stieff et al. 2011).

The relevance of abstractness becomes apparent in the case of phylogenetic trees (PTs) which are biology-specific representations of evolutionary relatedness. These representations underlie strict conventions and require a certain amount of content knowledge on evolution. Since PTs will be the basis of the following explanations we will shortly introduce this topic in the following section.

Excursus: The Nature of Phylogenetic Trees

PTs model evolutionary relationships among organisms. They serve both as models *of* existing hypotheses about those relationships as well as models *for* generating new hypotheses or testing existing hypotheses (Mahr 2008, 2009; Passmore et al. 2014; see "Representational Competence and Model Competence"). The representational forms of these models are manifold and students, teachers, and even scientists struggle with the correct reading, understanding, and interpreting of these representations (e.g., Baum et al. 2005; Halverson 2011). PTs appear in a variety of representational forms that reach from Darwin's first sketch over Haeckel's "pedigree of man" to modern phylograms and cladograms. As an example, cladograms show evolutionary relationships between recent taxa (groups of organisms like populations, species, and families) by depicting them as tips of the PT and their most recent common ancestors as nodes. In this special representational form branch length does not have a meaning (see Fig. 2). These examples adumbrate that good RC with PTs requires knowledge of several conventions. Being able to correctly read and interpret PTs is a crucial skill not only to deal with those repre-sentations but also to understand the concept of evolution (e.g., Baum et al. 2005; O'Hara, 1988). It is even stated that the way how individuals interpret PTs directly influences their understanding of evolution and vice versa (Omland et al. 2008). The consideration of life on earth in a phylogenetic manner and organization in PTs is defined as tree thinking and came up in the 1960s (O'Hara, 1988, 1997). Thus, it includes a hierarchical view on nature (Novick and Catley 2007), comprehension of evolutionary mechanisms, inheritance, and the use of phylogenetic tools such as PTs (Halverson et al. 2011). Hence, tree thinking can be referred to as a special, biology specific, kind of RC including not only the correct reading of PTs but also their interpretation and application.

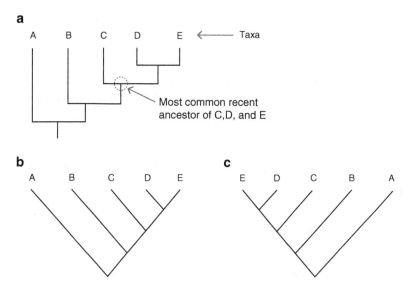

Fig. 2 Tree cladogram (**a**) and two ladder cladograms (**b, c**). All three cladograms represent the same evolutionary relationships

Visual Perception of Phylogenetic Trees

Novick et al. (2012) investigated the influence of ladderized cladograms' orientation on visual perception and RC. In this study, college students' understanding and reading direction of two different representations of ladder cladograms were compared: Cladograms with a "backbone"[7] from down left to up right (see Fig. 2b) versus top left to bottom right (see Fig. 2c). It was suggested that even though those cladograms represent the same information they are differently computed due to their representational form. The participants were college biology majors who had had at least one year of biology introduction. Eye movements were recorded using a head-mounted eye tracking system. Fixations and scanpaths revealed that students read ladder cladograms along the "backbone" and from left to right independent from its orientation. Additional data from two professors declared as experts showed a similar visual pattern (Novick et al. 2012). Due to the small sample size of experts, there is no profound comparison between experts and novices possible but at least eye tracking data revealed a certain trend. Subsequent to the eye tracking of ladder cladograms, students were shown tree cladograms (see Fig. 2a) which they had to compare to the ladder cladograms and decide whether they were identical with regards to content. They performed better with ladder cladograms oriented from top

---

[7] Ladder cladograms show a seemingly continuous line from the root to one taxon which is due to the high level of abstraction (see Figure 2.2 and 2.3). Students often have difficulties to understand the hierarchical character of ladder cladograms and rather perceive the "backbone" as a single line (Novick et al. 2012).

left to bottom right (see Fig. 2c) even though this form of ladder cladogram is less common than the other one. The authors of the study therefore suggested using this orientation when showing ladder cladograms (Novick et al. 2012). In summary, this study provides both data about scanning behavior when actually reading ladder cladograms and the tree reading abilities of the participants expressed in their performance in the transformation task. Thus the combination of two methods led to comprehensive insights into RC of students when dealing with PTs.

Representational Competence with Phylogenetic Trees – Levels and Milestones

The previously described study dealt with highly biology-specific representations and illustrates the content-specific character of RC. Even though the tested students in the study were biology-majors they struggled with less intuitively designed PTs. To explain students' difficulties, Halverson and Friedrichsen (2013) expanded Kozma and Russell's five levels of RC in chemistry (2005) to seven levels especially for RC in evolutionary contexts (see below) with PTs. The seven levels are defined as follows[8]:

1. None: Students do not recognize presented information.
2. Superficial: Students look at uninformative features like tip proximity and make no connection to underlying meaning.
3. Simplified: Students assume "main branch" with "side branches" and compare branches' similarities and differences between different representations.
4. Symbolic: Students understand symbolic elements but lay too much emphasis on nodes.
5. Conceptual: Students are able to rotate branches at nodes and see trees as 2D-illustrations of 3D-representations. They do no connect them to evolutionary history and are not able to compare trees with different styles.
6. Scientific: Students correctly read trees and are able to compare several representations independent from their style.
7. Expert: Multiple representations are used for scientific inquiry. This level is reserved for professionals like zoologists and does not distinguish between tree reading and building.

These levels were found to be dependent on the nature of tree representation and task, e.g., differed between a task on a single tree representation or comparison of multiple trees. The study comprised a one semester class on phylogenetic trees with pre and post testing of students' RC with trees. Trends showed most students being on level 2 to 3 prior to the class with improvement of around three levels at the end

---

[8] The study describes RC with PTs both for tree reading and tree building. Since this chapter focuses on visual aspects, it will only address tree reading. For further information about tree building, see Halverson and Friedrichsen (2013).

of the class. However this development was non-continuously and the level depended on the aspect of tree thinking (Halverson and Friedrichsen 2013). The seven levels build on milestones being essential for tree reading and building. Based on the skills that students need for RC in chemistry (Kozma and Russell 2005), Halverson and Friedrichsen (2013) could identify the following six milestones:

1. Recognition and interpretation of informative symbolic parts of a representation.
2. Comparison of multiple representations of similar nature to find the most appropriate one.
3. Accurate communication of a representation's meaning to others.
4. Predictions from a representation.
5. Testing and changing of representation according to new data.
6. Generating appropriate and accurate representations.

The first two milestones lead us back to eye tracking since visual perception of symbolic parts of a PT and the comparison of several representations are highly visual processes. The following section will address how these milestones and the seven levels could serve as theoretical background to assess RC with PTs using eye tracking technology.

Future Application – How to Assess Representational Competence with Phylogenetic Trees Using Eye Tracking Technology

RC with PTs has great potential being assessed via eye tracking since PTs combine complex representational forms with equally extensive underlying meaning and conventions as described above. PTs are a reasonable tool at school to teach evolution (Novick and Catley 2014) so future teachers should be competent tree thinkers to use this potential. Their level of RC with PTs could be diagnosed with the following sketch of a study basing on the first and second milestone for being a competent tree reader, combined with the seven levels of RC with PTs developed by Halverson and Friedrichsen (2013). The aim of the proposed study is to investigate differences in eye movements depending on the different levels of RC with PTs. Furthermore, this study diagnoses the RC level of participants in a comprehensive way by combining visual and verbal data.

Participants are pre-service biology teachers. They are at the end of being trained at school and university and will soon provide their knowledge to their students. Additionally, experts (e.g., systematic zoologists) are tested to get the "gold standard" of RC with PTs and to make sure that level 7 is met.

The idea is to present tasks with increasing difficulty to participants. Every task requires a certain level of RC to solve it. Hence, the first task requires RC level 1, second task RC level 2, and so on up to level 7. RC levels are known to be non-continuously (Halverson and Friedrichsen 2013; Kozma and Russell 2005) so every participant executes all seven tasks to ensure that the individual's maximum performance is reached. As long as participants are able to solve a task, they are considered to be on the tested level. Eye movements during task solving provide insights into which features of the representation participants use to solve the task.

The measures used in this study are fixations on AOIs relevant to solve the task (e.g., nodes, taxa), fixations on irrelevant AOIs (e.g., additional pictograms of taxa), duration of these fixations, time to first fixation of relevant AOIs, and order of fixations leading to scanpaths and possibly allows uncovering strategies. Tasks are presented on a stimulus screen and after each task verbal data are collected via cued RTA.

Differences between the seven levels could look as follows: Since participants on level 1 do not use the representations at all but rely on their prior knowledge, it can be assumed that they only fixate the tips to read the given species names. Probably, they will not pay attention to the actual informative part of the cladogram, namely the branching pattern and the nodes. Hence, they will rarely fixate nodes. Participants with level 2 RC skills (superficial use) will probably spend more time on the tips and the branches since they already try to extract information from the representation. Even though they are not able to correctly read a PT, they pay attention to the position of the tips and the branches. Hence, they would probably spend more time on the tips and go back and forth between the different taxa than participants on level 1. In contrast, participants on level 3 seem to focus on the branches which gives the opportunity to distinguish them from the other levels. Their focus on the "main line" can be investigated as Novick et al. (2012) already showed in their study with AOIs on the branches. The overemphasis of nodes is characteristic of level 4 (symbolic use) where students misleadingly count all nodes between species to determine relationships. Here, the order of fixations becomes relevant to distinguish purposeful fixations on nodes to count them from random node fixations. The task for level 5 includes mental rotation which is a characteristic skill on this level. Hence, participants have to mentally rotate branches at the nodes to compare two PTs with regard to similarities or differences. Here, participants might fixate the node at which they mentally rotate the branches longer than other nodes. At the scientific level 6, participants are able to compare different representational forms of PTs. It can be expected that participants fixate for example nodes in both representations in a target-oriented way to find differences and similarities. This can be revealed by scanpaths. Furthermore, it can be assumed that time to first fixation of relevant AOIs is shorter than for participants on lower levels.

Level 7 is a special case since it cannot be achieved by non-experts (Halverson and Friedrichsen 2013). However, since experts are tested, too, they are expected to reach this level by having their own and individual strategies (Jarodzka et al. 2013) to solve complex tasks including multiple representations in different representational forms.

Since the differences between the seven levels and especially between levels 1 to 4 might be hardly measurable via eye movements only, verbal data come into play. As discussed in "Complementary Data" cued RTA substantiates eye movement data and at the same time gives insights into why participants fixated a certain AOI (Stieff et al. 2011). These why-information could ease the discrimination between the levels.

An important point for data analysis is that it should only be performed for the maximum level a participant would reach. Since all participants run through all tasks from level 1 to 7, an expert solution of level 1 would differ from someone whose maximum is at level 1. As a consequence, eye movements and verbal data are compared between maximum levels, not between every level a participant performed.

Dependent on the data gained it might be possible that discrimination between the seven levels via eye movements is not selective enough. In this case a continuative idea would be to condensate some levels based on data to get down to merging levels like low, intermediate, or high.

One important control variable is the participants' level of model competence (Grünkorn et al. 2013; Upmeier zu Belzen and Krüger 2010). As described in "Representational Competence and Model Competence" and "Excursus: The Nature of Phylogenetic Trees", representations of PTs base on models *of* and *for* evolutionary relationships. Reading PTs and interpreting the underlying meaning, namely the model of evolutionary relationships, thus necessitates both RC and model competence. Therefore, when we investigate how individuals deal with representational forms of PTs, we should also illuminate how the underlying model is treated. This could lead to insights about if and how RC and model competence influence each other.

This proposed study does not only offer the possibility to diagnose RC with PTs using eye tracking technology. Expert eye movements during task solving can continuatively be used for eye movement modeling examples (EMME; van Gog et al. 2009) for enhancing learning with representations. For this approach an ideal scanpath (e.g., by an expert) is superimposed on the representation and replayed to show novices where relevant features are. This can either be done by a colored circle indicating fixation or a so called spotlight where only the area fixated by the expert is displayed in high solution whereas the rest of the representation is diffuse. Additionally, the expert's verbal report can simultaneously be played to give information about why the expert fixated a certain feature. Although EMME was shown to be detrimental in problem-solving tasks (van Gog et al. 2009; van Marlen et al. 2016), other studies have identified a positive impact on visual search and interpreting relevant features in description of fish locomotion patterns (Jarodzka et al. 2013) and clinical diagnosis (Jarodzka et al. 2012) as well as on integrative processing of text and graphics and transfer in reading an illustrated text (Mason et al. 2015). These latter three examples show that EMME would be suitable for instruction of tree reading to novices since this process includes finding relevant features of a representation and interpreting them.In sum, the proposed study exploits the potential of eye tracking technology combined with verbal data. First, it enables us to get insights into where individuals look at when reading PTs and why. Second, it might be possible to distinguish between levels of RC with PTs and therefore use eye tracking as tool for diagnosis. Third, the relation between RC and model competence could be highlighted. Fourth, expert solutions can be applied to teach tree reading to novices and thereby to improve their RC with PTs.

Conclusion

Visual perception is an important source of information for humans. Hence, it is no wonder that – with some exceptions – attention lies at the point we fixate with the eyes. Representations offer a lot of visual information so we need to distinguish between important and less important features to be able to interpret them. At this point, eye tracking technology comes into play as a powerful tool to assess RC. As described in this chapter, eye tracking offers a lot of opportunities to investigate how for example visual perception of representations of experts and novices differs and how we can benefit from these insights for instruction. Nevertheless, it might not be forgotten that data about eye movements on their own are not meaningful but need to be triangulated with verbal data in order to get information about the "why". Combining these information eye tracking enables us to connect learning outcomes to cognitive processes since we can investigate visual perception in an unfiltered way.

In conclusion, eye tracking technology is a promising tool to investigate visual perception processes when dealing with representations and thus to assess RC. Beyond that, it allows us to implement these insights to overcome drawbacks of representations for example in textbooks and to improve instruction.

References

Baum, D. A., DeWitt Smith, S., & Donovan, S. S. S. (2005). The tree-thinking challenge. *Science, 310*(5750), 979–980.

Boucheix, J.-M., & Lowe, R. K. (2010). An eye-tracking comparison of external pointing cues and internal continuous cues in learning with complex animations. *Learning and Instruction, 20*(2), 123–135.

Brinkman, J. A. (1993). Verbal protocol accuracy in fault diagnosis. *Ergonomics, 36*(11), 1381–1397.

Canham, M., & Hegarty, M. (2010). Effects of knowledge and display design on comprehension of complex graphics. *Learning and Instruction, 20*(2), 155–166.

Chi, M. T., Feltovich, P. J., & Glaser, R. (1981). Categorization and representation of physics problems by experts and novices. *Cognitive Science, 5*(2), 121–152.

Cook, M., Wiebe, E. N., & Carter, G. (2008). The influence of prior knowledge on viewing and interpreting graphics with macroscopic and molecular representations. *Science Education, 92*(5), 848–867.

Duchowski, A. T. (2002). A breadth-first survey of eye-tracking applications. *Behavior Research Methods, Instruments, & Computers, 34*(4), 455–470.

Duchowski, A. T. (2007). *Eye tracking methodology: Theory and practice* (2nd ed.). London: Springer. With permission of Springer.

Ericsson, K. A., & Kintsch, W. (1995). Long-term working memory. *Psychological Review, 102*, 211–245.

Ericsson, K. A., & Lehmann, A. C. (1996). Expert and exceptional performance: Evidence of maximal adaptations to task constraints. *Annual Review of Psychology, 47*, 273–305.

Ericsson, K. A., & Simon, H. A. (1993). *Protocol analysis: Verbal reports as data* (Rev. ed.). Cambridge, MA: MIT Press.

Feusner, M., & Lukoff, B. (2008). Testing for statistically significant differences between groups of scan patterns. In S. N. Spencer (Ed.), *Proceedings of the 2008 symposium on eye tracking research & applications* (pp. 43–46). New York: ACM.

Gegenfurtner, A., Lehtinen, E., & Säljö, R. (2011). Expertise differences in the comprehension of visualizations: A meta-analysis of eye-tracking research in professional domains. *Educational Psychology Review, 23*(4), 523–552.

Gilbert, S. W. (1991). Model building and a definition of science. *Journal of Research in Science Teaching, 28*(1), 73–79.

van Gog, T., Paas, F., Merriënboer, v., Jeroen, J. G., & Witte, P. (2005). Uncovering the problem-solving process: Cued retrospective reporting versus concurrent and retrospective reporting. *Journal of Experimental Psychology. Applied, 11*(4), 237–244.

van Gog, T., Jarodzka, H., Scheiter, K., Gerjets, P., & Paas, F. (2009). Attention guidance during example study via the model's eye movements. *Computers in Human Behavior, 25*(3), 785–791.

Grünkorn, J., Upmeier zu Belzen, A., & Krüger, D. (2013). Assessing Students' understandings of biological models and their use in science to evaluate a theoretical framework. *International Journal of Science Education, 36*(10), 1651–1684.

Haider, H., & Frensch, P. A. (1999). Eye movement during skill acquisition: More evidence for the information reduction hypothesis. *Journal of Experimental Psychology: Learning, Memory, & Cognition, 25*, 172–190.

Halverson, K. L. (2011). Improving tree-thinking one learnable skill at a time. *Evolution: Education and Outreach, 4*(1), 95–106.

Halverson, K. L., & Friedrichsen, P. (2013). Learning tree thinking: Developing a new framework of representational competence. In D. F. Treagust & C.-Y. Tsui (Eds.), *Models and modeling in science education: Vol. 7, Multiple Representations in Biological Education* (pp. 185–201). Dordrecht: Springer.

Halverson, K. L., Pires, C. J., & Abell, S. K. (2011). Exploring the complexity of tree thinking expertise in an undergraduate systematics course. *Science Education, 95*(5), 794–823.

Holmqvist, K., Nyström, M., Andersson, R., Dewhurst, R., Jarodzka, H., & van de Weijer, J. (2011). *Eye tracking: A comprehensive guide to methods and measures*. Oxford, New York: Oxford University Press.

Jarodzka, H., Scheiter, K., Gerjets, P., & van Gog, T. (2010). In the eyes of the beholder: How experts and novices interpret dynamic stimuli. *Learning and Instruction, 20*(2), 146–154.

Jarodzka, H., Balslev, T., Holmqvist, K., Nyström, M., Scheiter, K., Gerjets, P., & Eika, B. (2012). Conveying clinical reasoning based on visual observation via eye-movement modeling examples. *Instructional Science, 40*(5), 813–827.

Jarodzka, H., van Gog, T., Dorr, M., Scheiter, K., & Gerjets, P. (2013). Learning to see: Guiding students' attention via a Model's eye movements fosters learning. *Learning and Instruction, 25*, 62–70.

Just, M. A., & Carpenter, P. A. (1980). A theory of reading: From eye fixations to comprehension. *Psychological Review, 87*, 329–354.

Koning, d., Björn, B., Tabbers, H. K., Rikers, R. M., & Paas, F. (2010). Attention guidance in learning from a complex animation: Seeing is understanding? *Learning and Instruction, 20*(2), 111–122.

Kozma, R., & Russell, J. (1997). Multimedia and understanding: Expert and novice responses to different representations of chemical phenomena. *Journal of Research in Science Teaching, 34*(9), 949–968.

Kozma, R., & Russell, J. (2005). Students becoming chemists: Developing representational competence. In J. K. Gilbert (Ed.), *Visualization in science education* (pp. 121–145). Dordrecht: Springer.

Kozma, R., Chin, E., Russell, J., & Marx, N. (2000). The roles of representations and tools in the chemistry laboratory and their implications for chemistry learning. *The Journal of the Learning Sciences, 9*(2), 105–143.

Kundel, H. L., Nodine, C. F., Conant, E. F., & Weinstein, S. P. (2007). Holistic component of image perception in mammogram interpretation: Gaze-tracking study. *Radiology, 242,* 396–402.

Lai, M.-L., Tsai, M.-J., Yang, F.-Y., Hsu, C.-Y., Liu, T.-C., Lee, S. W.-Y., et al. (2013). A review of using eye-tracking technology in exploring learning from 2000 to 2012. *Educational Research Review, 10,* 90–115.

Mahr, B. (2008). Ein Modell des Modellseins: Ein Beitrag zur Aufklärung des Modellbegriffs. In E. Knobloch & U. Dirks (Eds.), *Modelle* (pp. 187–218). Frankfurt am Main: Peter Lang.

Mahr, B. (2009). Die Informatik und die Logik der Modelle. *Informatik Spektrum, 32*(3), 228–249.

van Marlen, T., van Wermeskerken, M., Jarodzka, H., & van Gog, T. (2016). Showing a model's eye movements in examples does not improve learning of problem-solving tasks. *Computers in Human Behavior, 65,* 448–459.

Mason, L., Pluchino, P., & Tornatora, M. C. (2015). Eye-movement modeling of integrative reading of an illustrated text: Effects on processing and learning. *Contemporary Educational Psychology, 41,* 172–187.

McMains, S. A., & Kastner, S. (2009). Visual Attention. In M. D. Binder, N. Hirokawa, & U. Windhorst (Eds.), *Encyclopedia of neuroscience* (pp. 4296–4302). Berlin, Heidelberg: Springer.

Nersessian, N. J. (2002). The cognitive basis of model-based reasoning in science. In *The cognitive basis of science* (pp. 133–153).

Nitz, S., Ainsworth, S. E., Nerdel, C., & Prechtl, H. (2014). Do student perceptions of teaching predict the development of representational competence and biological knowledge? *Learning and Instruction, 31,* 13–22.

Novick, L. R., & Catley, K. M. (2007). Understanding phylogenies in biology: The influence of a gestalt perceptual principle. *Journal of Experimental Psychology: Applied, 13*(4), 197–223.

Novick, L. R., & Catley, K. M. (2014). When relationships depicted diagrammatically conflict with prior knowledge: An investigation of students' interpretations of evolutionary trees. *Science Education, 98*(2), 269–304.

Novick, L. R., Stull, A. T., & Catley, K. M. (2012). Reading phylogenetic trees: The effects of tree orientation and text processing on comprehension. *Bioscience, 62*(8), 757–764.

O'Hara, R. J. (1988). Homage to Clio, or, toward an historical philosophy for evolutionary biology. *Systematic Zoology, 37*(2), 142–155.

O'Hara, R. J. (1997). Population thinking and tree thinking in systematics. *Zoologica Scripta, 26*(4), 323–329.

Omland, K. E., Cook, L. G., & Crisp, M. D. (2008). Tree thinking for all biology: The problem with reading phylogenies as ladders of progress. *BioEssays, 30*(9), 854–867.

Passmore, C., Gouvea, J. S., & Giere, R. (2014). Models in science and in learning science: Focusing scientific practice on sense-making. In M. R. Matthews (Ed.), *International handbook of research in history, philosophy and science teaching* (1st ed., pp. 1171–1202). Dordredht: Springer Netherlands.

Rayner, K. (1998). Eye movements in reading and information processing. 20 years of research. *Psychological Bulletin, 124*(3), 372–422.

Rayner, K., Pollatsek, A., Ashby, J., & Clifton, C., Jr. (2012). *Psychology of reading. New York*: Psychology Press.

van Someren, M. W., Barnard, Y. F., & Sandberg, J. A. C. (1994). *The think aloud method: A practical guide to modeling cognitive processes*. London: Academic Press.

Stieff, M. (2007). Mental rotation and diagrammatic reasoning in science. *Learning and Instruction, 17*(2), 219–234.

Stieff, M., Hegarty, M., & Deslongchamps, G. (2011). Identifying representational competence with multi-representational displays. *Cognition and Instruction, 29*(1), 123–145.
Tippett, C. D., & Yore, L. (2011). Exploring middle school students' representational competence in science: Development and verification of a framework for learning with visual representations.
Upmeier zu Belzen, A., & Krüger, D. (2010). Modellkompetenz im Biologieunterricht. *ZfDN, 16,* 41–57.

The Use of a Representational Triplet Model as the Basis for the Evaluation of Students' Representational Competence

Jill D. Maroo and Sara L. Johnson

Introduction

This chapter addresses how to use a representational triplet model to evaluate students' representational competence. We agree with the detailed description of representational competence found in "Towards a Definition of Representational Competence". Namely, that representational competence is an individual's ability to use representations to explain content knowledge. We view content knowledge broadly, as all the knowledge an individual displays which pertains to a specific topic of study. While the topic of this book is representational competence, student understanding is at the core of our research interest.

Student understanding has a variety of valid, but diverse, definitions. For this chapter, we consider student understanding a combination of an individual's content knowledge and representational competence. We consider these two entities distinct but related based on our previous research (Johnson et al. 2010). In the work we present here, we first investigated representational competence independent of content knowledge accuracy. We then accounted for content knowledge accuracy and found the representational competence for the majority of students was stable. Using a representational triplet model to investigate representational competence, in this way, allowed us to obtain a more detailed picture of student understanding.

J. D. Maroo (✉)
University of Northern Iowa, Cedar Falls, IA, USA
e-mail: jill.maroo@uni.edu

S. L. Johnson
University of North Alabama, Florence, AL, USA
e-mail: sjohnson34@una.edu

© Springer International Publishing AG, part of Springer Nature 2018
K. L. Daniel (ed.), *Towards a Framework for Representational Competence in Science Education*, Models and Modeling in Science Education 11,
https://doi.org/10.1007/978-3-319-89945-9_12

Triplet Models

Triplet models exist throughout chemistry education research, although their application has expanded into other disciplines (Talanquer 2011). Triplet models organize knowledge into three levels; connections can be made between these levels to help explain or describe the natural world. While the organization of knowledge into levels remains consistent, the names, definitions, and number of levels vary (Talanquer 2011; Gilbert and Treagust 2009).

The organization of chemical knowledge into a triplet model finds its roots in Johnstone's (1982) early work. Johnstone's initial organization contained three levels: description and functional, representational, and explanatory. However, within his early model, Johnstone included the separation of macrochemistry from microchemistry. This eventually developed into what we call Johnstone's triangle, a triplet model containing three levels: macroscopic, sub-microscopic and either symbolic or representational (Johnstone 1991, 1993). In several of Johnstone's (1982, 1991, 1993) works, he emphasizes the equality of the three levels but explains that all individuals do not need to use all levels. Through these works, we observe Johnstone using a variety of terminology to describe what his models are communicating. Similarly, we see this trend continue in research that uses triplet models.

Talanquer (2011) states that the organization of chemical knowledge into three levels is one of the most prolific models for chemical education research. Many researchers have used the triplet model lens to guide their data analysis (e.g. Boddey and de Berg 2014; Rappoport and Ashkenazi 2008; Treagust et al. 2003). This lens has further been useful in the design of assessment tools (e.g. Chandrasegaran et al. 2007; Jaber and BouJaoude 2012). While the examples listed are all within chemical education, research applying the triplet model also exists in other educational disciplines, specifically biology (e.g. Marbach-Ad and Stavy 2000).

Our Triplet Model

We believe the representational nature of our triplet model makes it a valuable tool for discussing representational competence. Our model provides a scaffold for constructing a representation of an individual's ability to work with conceptual knowledge. This representation paints an overall picture of their representational competence. We designate our triple model levels as: representational macroscopic, representational microscopic and representational symbolic (see Fig. 1). We consider the representational macroscopic level as anything visible to a student with average visual acuity, including all interactions occurring at that level. The representational microscopic level is anything not visible to a student with average visual acuity. The representational symbolic level is any tool (e.g. a symbol, formula, or diagram) used to explain the representational macroscopic or representational microscopic levels.

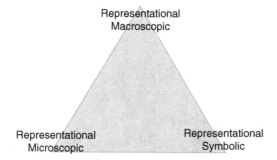

Fig. 1 Visual representation of the triplet model used in this study. Representational level names are located on the points of the triangle. Lines connecting the points represent the highest potential connections individuals can make between levels. Individuals with expert representational competence are able to connect all three points and explore the center of the triangle

To demonstrate, we will use our triplet relationship to describe how we think about the phenomenon of dissolving table sugar into water. When we talk about dissolving table sugar into water, we may choose to represent this process using symbols (i.e. representational symbolic level). For example, we can represent solid table sugar using a square, while representing water with an alternative shape of a circle. To represent dissolved sugar, we may choose to use a third shape (e.g. triangle) or combine our square and circle in some way to demonstrate the interaction between water and sugar. On the representational macroscopic level, we can demonstrate the phenomenon of sugar dissolving by adding a visible quantity of table sugar to a beaker filled with a suitable amount of water. As we stir the solution, the table sugar will dissolve and reduce in quantity, until it seemingly disappears from our vision. Alternatively, we can use a low power microscope to watch the same process on the representational microscopic level. On this level, we can observe how the sugar crystal's edges diffuse and its overall size reduces as the sugar dissolves into the water. In addition to these observations, we understand that table sugar dissolves in water because of its ability to interact with water via non-covalent interactions. While these interactions cannot be seen, even with a microscope, we will consider this as representational microscopic, as defined by our triplet model terms.

Analyzing Data with Our Triplet Model

To demonstrate one way to analyze data using our triplet model, we present data from a previous study on alternative conceptions (Johnson et al. 2010). Our qualitative study explored students' alternative conceptions related to the Tricarboxylic Acid (TCA) cycle using semi-structured interviews (see Fig. 2). We generated a taxonomy through inductive coding of student interview excerpts. Our concept taxonomy was informed by the alternative conceptions demonstrated by undergraduate students in an introductory biology class.

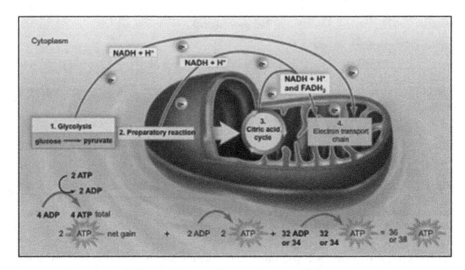

Fig. 2 The TCA cycle is also referred to as the Krebs cycle or the Citric Acid cycle. A simplified description of the TCA cycle follows. The TCA cycle is the component of cellular respiration that links glycolysis to the electron transport chain (ETC), ultimately resulting in the production of adenosine triphosphate (ATP), the cell's energy currency. In eukaryotes, the three carbon product of glycolysis, pyruvate, moves from the cytosol into the mitochondrial matrix where it is converted to the two carbon molecule, acetyl-CoA. Acetyl-CoA enters the cycle by condensing with the oxaloacetate to form the six carbon molecule, citrate. Citrate enters a series of reversible reactions producing electron donating compounds (e.g. NADH) that are then used in the ETC to power oxidative phosphorylation and make ATP. In addition to its role in carbohydrate metabolism, the TCA cycle integrates other metabolic processes in the cell. This interconnected nature requires the TCA cycle be dynamic and highly regulated

Taxonomy of Concepts Related to the TCA Cycle

- Identification of mitochondrion representation
- Identification of the meaning of the mitochondrion's moniker "powerhouse"
- TCA cycle's connection to photosynthesis
- TCA cycle's connection to fermentation
- Relative placement of glycolysis in metabolism
- Relative placement of electron transport chain (ETC) in metabolism
- Role of oxygen connected to TCA cycle
- Number of substrates entering TCA cycle at one time
- Number of TCA cycle enzymes in one mitochondrion
- Dynamic nature of TCA cycle (Johnson et al. 2010)

We further explored student representational competence by analyzing these excerpts with our triplet model. The results of the triplet model analysis provided us with visual tools to communicate the representational competence of either an individual or a group. We refer back to the lettered list above when demonstrating our analysis.

Individual Representational Competence

When investigating individual representational competence, we defined one excerpt as all discussion an individual had on one specific concept during an interview. For most, these discussions were not continuous but were distributed throughout the interview. Our taxonomy contained ten concepts informed by alternative conceptions related to the TCA cycle (Johnson et al. 2010). We coded each excerpt using our triplet model. We then differentiated between correct and incorrect discussion for each concept, yielding a maximum of 20 possible excerpts for each student. Each dot on an individual's dot plot represents the highest representational competence exhibited for that excerpt.

To code excerpts, we identified whether the individual used representational microscopic, representational macroscopic, representational symbolic or a combination of language to discuss each concept. If an individual used more than one level of our triplet model (e.g. representational microscopic and representational macroscopic), we then identified whether the excerpt made connections between the levels or treated the triplet model levels as individual entities. If the excerpt did treat the triplet model levels as individual entities, we coded the excerpt as individual points. This identification led to the creation of hierarchical categories ranging from a low of one point through a high of connecting all points and exploring the center of the triangle. The categories included: one point, two points, three points, one connection, one connection and one point, two connections, three connections and exploring the center.

The following quote is from Dillon, who discussed a concept (i.e. the presence of many cycles within a cell) on two levels, but did not make connections between the levels. He said, "I think **it happens** [the net gain of 2 ATP], all the way, everywhere...there's an *ATP* synthesis pump." Because Dillon made no connections between levels, we coded his excerpt as two unconnected points. We have indicated the triplet model levels within the text.

As seen above (in bold), Dillon referenced the representational microscopic level when giving a reason for why more than one cycle occurred simultaneously within a cell. We coded Dillon's response at the representational symbolic (in italics) because of his use of the abbreviation of adenosine triphosphate, ATP. We considered these levels as unconnected points because Dillon did not make an explicit reference to how the ATP synthesis pump related to his microscopic level explanation (the net gain of 2 ATP).

The quote below is from Ada, who discussed a concept related to glycolysis on three levels. Ada made two different connections between two levels, but did not connect the third pair. In addition to the codes employed above, we use underlining to indicate the representational macroscopic level.

So, the TCA cycle. You eat your food and it would break down. **Glycolysis** where all the sugars come apart. And then you'd have **glucose** and...what is it? [sorting through cards] So, then you'd have *ATP* comes out at the end. But I can't remember if. I know that some cycle produces *ATP* but then it uses the *ATP* to create more *ATP*, so I don't remember if that's this or not. I don't know if it'd be in there. But I believe that it kinda goes like that. Cause I know. I think the **hydrogen** comes off of that. Then you've got the *ADP*. And then you'd have H_2O, but I don't know what happened to the **carbon**.

We coded Ada's excerpt as having connections between the representational macroscopic and microscopic levels, and between the representational symbolic and microscopic levels. Ada did not make connections between the representational symbolic and macroscopic levels.

At the beginning of the excerpt, Ada engaged the representational macroscopic level with the phrase "you eat your food and it would break down." She then restates and expands her idea to include the representational microscopic level, explaining, "glycolysis where all the sugars come apart." With her connection between two levels, she demonstrates her understanding of how the body uses food. At the end of the excerpt, Ada began to describe her understanding of the TCA cycle. In her explanation, she uses language across the representational symbolic and microscopic levels. She is able to elevate her explanation beyond terminology, creating a relationship component.

After coding our excerpts, we identified the highest triplet model category reached. We developed a dot plot for each individual where each dot represented one excerpt. Because dot plots depended on individual interviews, the number of excerpts (dots) on the dot plots varied. These dot plots provided an overall picture of an individual's representational competence. This qualitative analysis included two researchers deliberating the trend of each dot plot until we reached agreement on individual's placement on the triplet category continuum. As an example, we have provided the dot plot below containing 20 excerpts without regard to accuracy (see Fig. 3).

The dot plot in Fig. 3 gives an overall picture of an individual's representational competence, one component of student understanding. However, if we add a component of content knowledge (i.e. accuracy), we can begin to explore a student's understanding. We have provided an example dot plot that takes into account content accuracy in order to obtain a measure of student understanding on a specific topic (see Fig. 4).

Robert's representational competence dot plot appears below in Fig. 5. We classified Robert as a student with low representational competence. When looking at Robert's dot plot, we observed the overall shape to determine his representational competence on this topic. We identified two of Robert's excerpts as one-connection using our triplet categories. However, his dot plot was heavily skewed toward the lower range with 75% of the dots in the one point category. We considered Robert's two one-connection dots to increase his overall representational competence slightly. Therefore, we described Robert's overall representational competence as falling between the one and two point categories.

Figure 6 illustrates the representational competence dot plot for Ada, a student with medium representational competence. We observed the overall shape of Ada's dot plot to determine her representational competence on this topic. Although Ada had a similar number of excerpts as Robert, her dot plot shape was noticeably

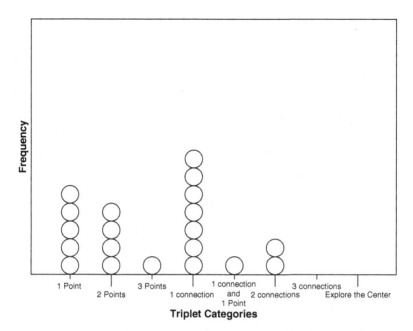

Fig. 3 Example of hypothetical individual dot plot. Dot plot contains 20 excerpts (dots) arranged by their triplet category. Excerpt accuracy is not indicated on the dot plot. This individual's overall placement on the triplet category continuum is between three points and one connection categories

different. We classified over half of Ada's excerpts as within the medium representational competence range. While Ada's dot plot contained four dots in the low range, she only had one one-point dot and her highest two dots appeared at the two-connection triplet category. Therefore, when determining her representational competence, we considered the locations of all of her dots. We classified her overall representational competence as falling between the three points and one-connection categories.

In our example, we investigated individuals' representational competences by constructing dot plot representations from coded interview excerpts. We highlighted our analysis for measuring individual representational competence in this section. We constructed individual dot plots using this approach for all of our participants with regards to accuracy (see Fig. 7). Using these plots, we then extended our approach to depict group representational competence.

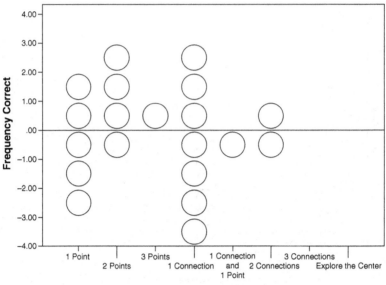

Fig. 4 Example of hypothetical individual dot plot depicting accuracy. Dot plot contains 20 excerpts (dots) arranged by their triplet category. Accuracy is indicated on the dot plot, with correct excerpts positioned above the line and incorrect excerpts positioned below the horizontal line. This individual's overall placement on the triplet category continuum is slightly above the three points category

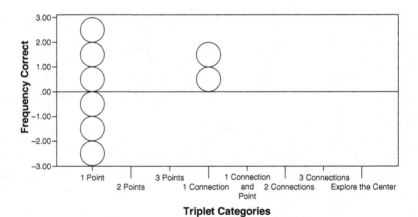

Fig. 5 Individual dot plot depicting accuracy for Robert, a student with low representational competence

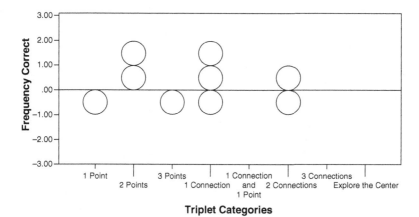

Fig. 6 Individual dot plot depicting accuracy for Ada, a student with medium representational competence

Group Representational Competence

We created a spectrum to visually display group representational competence (see Fig. 8) guided by our data and previous literature (Kozma and Russell 2007). Individuals in the low range had an overall representational competence displaying no connections between levels in our triplet model. Medium range individuals had overall representational competences that placed them between three points up to two connections between levels in our triplet model. For individuals to place in a high range, their overall representational competence fell above two connections to three connections between levels in our triplet model. Expert placement required an overall representational competence with above three connections to exploring the middle of our triplet model triangle.

We compiled our assessment of each individual's representational competence (Table 1) to construct a spectrum displaying the group representational competence of our participants. We visually assessed each dot plot in order to determine an overall representational competence value for an individual. The data for this spectrum comes from the analysis of multiple individual representational competence dot plots as detailed in the examples above. Our assessment included both the correct and incorrect regions of the dot plot. We plotted these representational competence values on our spectrum (see Fig. 8). This spectrum helped to identify our group's range of representational competence as low to medium. The spectrum also allowed us to observe that our overall group contained individuals in two different locations: low and medium (see Fig. 9). Our findings indicate that all students had a representational competence with an average below two connections between levels in our triplet model. This is not surprising since all individuals in our study were undergraduate students.

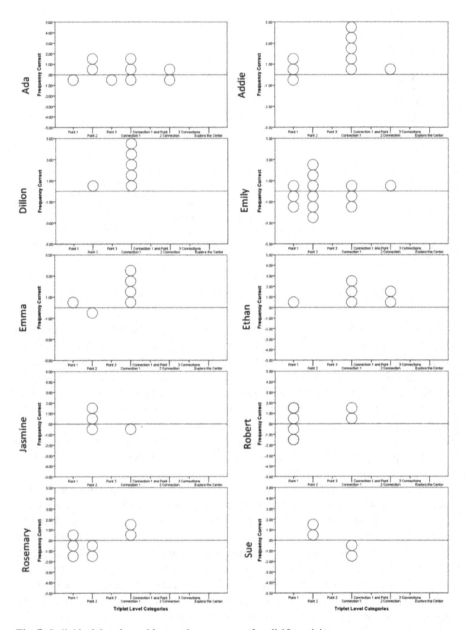

Fig. 7 Individual dot plots with regards to accuracy for all 10 participants

To determine how accuracy influenced our group's overall representational competence spectrum, we reassessed their dot plots using only the correct regions. We plotted these representational competence values on a new spectrum (see Fig. 10). From the correct responses spectrum, we see the range of representational compe-

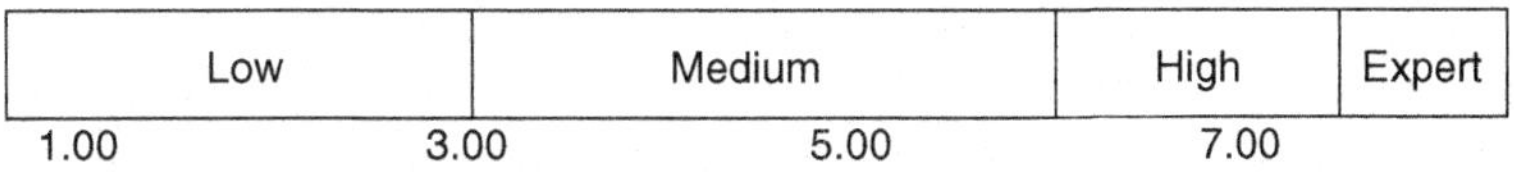

Low	Medium	High	Expert

1.00   3.00   5.00   7.00

Fig. 8 Example representational competence spectrum without student data. Image shows the range of four representational competence groups on the spectrum

Table 1 Individual placement on the representational competence spectrum

Individual	Without regards to accuracy	With regards to accuracy
Ada	Medium	Medium
Addie	Medium	Medium
Dillon	Medium	Medium
Emily	Low	Low
Emma	Medium	Medium
Ethan	Medium	Medium
Jasmine	Low	Low
Robert	Low	Low
Rosemary	Low	Low/medium*
Sue	Low/medium*	Medium

*Individuals placed at the upper range limit were counted within the lower range.

Fig. 9 Representational competence spectrum with student data from all responses. Each diamond represents an individual student's representational competence dot plot. Difference in fill indicates the two different spectrum locations of representational competence seen in this study

Fig. 10 Representational competence spectrum with student data from correct responses. Each diamond represents an individual student's representational competence dot plot. Eight diamonds are visible because there were two incidents of identical representational competence values. Difference in fill indicates the two different spectrum locations of representational competence seen in this study

tence values is similar to Fig. 9. While some individual's placement on the spectrum changed, the overall distribution and pattern of individuals into the two locations remained consistent. One student moved from the upper limit of the low representational competence group to the medium representational competence group. The spectrum appears to be missing two members because of two incidents of identical representational competence in the medium group. These findings lend evidence to the argument that representational competence and content knowledge are separate components of student understanding.

In our example, we used a representational competence spectrum to investigate a group of students' overall representational competence concerning the TCA cycle at one point in time. However, our approach can be modified to compare multiple groups of interest at one point or one group of interest across time. Teachers and researchers can use this approach to obtain a broader picture of their students' understanding.

Guidelines for Assessment

We consider student understanding a combination of an individual's content knowledge and representational competence. Although these guidelines focus on representational competence, we cannot ignore its association with content knowledge in student understanding. However, we argue that representational competence is not fully dependent on correct content knowledge. For researchers and teachers to gain a full picture of students' understanding, they must consider assessments that measure both.

Student understanding is of interest to both researchers and teachers. In this section, we combine our discussion into topics of relevance across the two groups. While the motivations of these two groups may be different, there are many similarities in the foundational steps of assessment design. The steps we discuss in this section include: question (item) development, assessment (instrument) development, and interpretation of results.

Question Development

It stands to reason that one of the greatest influences on the answers you receive from students is the question you ask. We can consider the components of a question through many lenses. For example, we can analyze a question using Bloom's taxonomy (Bloom et al. 1956) to determine the cognitive domain required to give a complete answer. Similarly, you can use our representational triplet model to classify questions based on the triplet categories. To do this, you must consider the two overarching components present in our triplet categories: the individual representational levels present and any connections between those levels. These components may exist in both question text and more importantly your expected answers.

We suggest you write questions that require a variety of answers across levels and connections between the levels. Some questions should only require answers on one of the triplet level categories (representational microscopic, representational symbolic, or representational macroscopic). Your question bank should contain questions that focus on each of the individual categories. Other questions should require answers on multiple levels without necessarily calling for connections between levels. Questions expecting higher order representational competence require answers to connect a minimum of two triplet categories. Similar to Bloom's Taxonomy (Bloom et al. 1956) creating questions over a variety of levels will allow you to assess your students' representational competences. Our suggestions may apply to questions for students at all levels; however, we acknowledge our study did not include students at levels other than undergraduate.

Assessment Development

When developing an assessment you should pull questions that cover a variety of levels and connections in their answers. This variety ensures that students at different representational competence levels have a chance to correctly answer the question according to the triplet model. However, students with higher representational competence compared to their peers may give you higher triplet category levels than you requested or expected. As previously stated, students' answers can cover the correct triplet model category but display incorrect content. We suggest building the assessment to increase in triplet category complexity starting with the lowest complexity and ending with your highest level.

A complete assessment does not have to contain questions on all triplet category levels. Similar to Bloom's Taxonomy, you should be mindful of the level for which you are writing the assessment. There is a strong possibility your students cannot reach the high or expert levels during an assessment, depending on the topic. We found students in an undergraduate introductory biology class did not display abilities above the category of two connections. Your highest triplet category assessed should be slightly above what you expect is the representational competence of your audience.

Interpretation of Results

The true measure for a student's representational competence comes when you score your assessment using the approaches given in the analyzing data section. When scoring for representational competence, keep in mind your questions may influence your students' answers. We noticed in our research many of our questions focused on the microscopic level of the triplet model. This may have influenced the extent of student's ability to make connections between the levels on their own. We caution you to be mindful of this limitation when interpreting the results of either dot plots or spectra.

Our method of assessment helps identify multiple types of students when you observe their representational competence compared to their content knowledge on a topic. First, you can determine which students have a higher representational competence, but lack understanding of the content. Second, you can identify the students who struggle with representational competence, although they have correct content knowledge. You can also observe, through dot plots, which students are either low or high in both representational competence and content knowledge. This paints an overall picture of a student's knowledge, which is useful to compare or track students.

Checklist

The following is a checklist that synthesizes our guidelines and presents them in an abbreviated form. When using our triplet model to assess representational competence you should:

1. Create questions that cover a variety of triplet categories.
2. Indicate explicitly which triplet category is necessary to answer the question.
3. Identify the triplet category needed to give a fully correct answer to your question.
4. Frame your assessment around a variety of questions.
5. Maintain reasonable expectations when assessing the representational competence your students possess.
6. Use multiple assessments to gain a complete picture of a student's representational competence

Potential Uses

Educators will find this assessment useful to gain a broad picture of their class's representational competence. They may also find it useful to look at an individual student's representational competence before and after instruction. Although we believe an individual's representational competence may be content dependent, it stands to reason that individuals will have similar measurements for topics with which they are unfamiliar. A further use of this assessment is to obtain a detailed picture of each student's understanding on the topic, when you separate correct and incorrect content responses. Action researchers may find these same methods helpful to continually assess student and class representational competences.

In addition to the classroom uses, these methods are applicable to research settings. Depending on your research question, you may use this model to obtain an individual's representational competence, a group's representational competence spectrum, or a measure of an individual's overall understanding on a topic. You can modify any of these approaches to answer a variety of research questions, both comparative and descriptive in nature.

Conclusions

Our chapter advocates the use of a representational triplet model for exploring student understanding. Herein, we have described how we used this tool to measure representational competence and student understanding for individuals. We further presented a way to compare individuals' representational competences as a group or across time. Based on our experiences and current research, we have presented guidelines for assessment design based on this triplet model. We want to reiterate the importance of aligning your assessment to the triplet model categories in order to achieve your desired outcomes. This approach has implications in both educational and research settings.

References

Bloom, B. S., Engelhart, M. D., Furst, E. J., Hill, W. H., & Krathwohl, D. R. (1956). In B. S. Bloom (Ed.), *Taxonomy of educational objectives: The classification of educational goals, Handbook 1: Cognitive domain*. New York: David McKay Company, Inc.

Boddey, K., & de Berg, K. (2014). The impact of nursing students' prior chemistry experience on academic performance and perception of relevance in a health science course. *Chemistry Education Research and Practice*. Advance online publication, *16*, 212. https://doi.org/10.1039/c4rp00240g.

Chandrasegaran, A. L., Treagust, D. F., & Mocerino, M. (2007). The development of a two-tier multiple-choice diagnostic instrument for evaluating secondary school students' ability to describe and explain chemical reactions using multiple levels of representation. *Chemistry Education Research and Practice, 8*(3), 292–307. https://doi.org/10.1039/B7RP90006F.

Gilbert, J. K., & Treagust, D. F. (2009). Introduction: Macro, submicro and symbolic representations and the relationship between them: Key models in chemical education. In J. K. Gilbert & D. F. Treagust (Eds.), *Models and modeling in science education, Volume 4: Multiple representations in chemical education* (pp. 1–8). Netherlands: Springer.

Jaber, L. Z., & BouJaoude, S. (2012). A macro-micro-symbolic teaching to promote relational understanding of chemical reactions. *International Journal of Science Education, 34*, 973–998. https://doi.org/10.1080/09500693.2011.569959.

Johnson, S. L., Maroo, J. D. & Halverson, K. L. (2010, April). Classification of UndergraduateAlternative Conceptions of the Tricarboxylic Acid Cycle. Paper presented at the meeting of National Association for Research in Science Teaching, Orlando, FL.

Johnstone, A. H. (1982). Macro- and microchemistry. *School Science Review, 64*, 377–379.

Johnstone, A. H. (1991). Why is science difficult to learn? Things are seldom what they seem. *Journal of Computer Assisted Learning, 7*, 75–83. https://doi.org/10.1111/j.1365-2729.1991.tb00230.x.

Johnstone, A. H. (1993). The development of chemistry teaching: A changing response to changing demand. *Journal of Chemical Education, 70*, 701–705. https://doi.org/10.1021/ed070p701.

Kozma, R., & Russell, J. (2007). Students becoming chemists: Developing representational competence. In J. Gilbert (Ed.), *Visualization in science education* (pp. 121–145). Dordrecht: Springer.

Marbach-Ad, G., & Stavy, R. (2000). Students' cellular and molecular explanations of genetic phenomena. *Journal of Biological Education, 34*(4), 200–205. https://doi.org/10.1080/00219266.2000.9655718.

Rappoport, L. T., & Ashkenazi, G. (2008). Connecting levels of represenation: Emergent versus submergent perspective. *Internation Journal of Science Education, 30*, 1585–1603. https://doi.org/10.1080/09500690701447405.

Talanquer, V. (2011). Macro, submicro, and symbolic: The many faces of the chemistry "triplet". *International Journal of Science Education, 32*, 1–17. https://doi.org/10.1080/09500690903386435.

Treagust, D., Chittleborough, G., & Mamiala, T. (2003). The role of submicroscopic and symbolic representations in chemical explanations. *International Journal of Science Education, 25*, 1353–1368. https://doi.org/10.1080/0950069032000070306.

Representational Competence in Science Education: From Theory to Assessment

Jochen Scheid, Andreas Müller, Rosa Hettmannsperger, and Wolfgang Schnotz

Theory

For a proper understanding of scientific experiments, phenomena and concepts, various mental representations, and their interplay are vitally important (Gilbert and Treagust 2009; Mayer 2005). The term 'representation' is understood as a tripartite relation of a referent (or object), its representation, and the meaning (or interpretation) of referent, representation and of their interaction. This relation is referred to in various ways (e.g. 'Peircean triangle', or 'triangle of meaning'), and a detailed discussion of its underpinnings in epistemology and semiotics can be found in Tytler et al. (2013, Ch. 6.). Schnotz (in line with dual coding theory) distinguishes two types of representations (2005, 2002, Integrated Text and Picture Comprehension). Photographs or schematic drawings are depictive representations. Formulas, tables, and verbal descriptions are descriptive representations. All forms of representation can appear internally (in the mind) or externally (e.g., on a paper or screen). Representations are selective; therefore they can differ in content and can be useful for solving different tasks (see Schnotz 1994; Herrmann 1993). A representation is not necessarily only an illustrative picture. In addition, it can be

J. Scheid (✉) · W. Schnotz
University of Koblenz-Landau, Landau, Germany
e-mail: scheid@uni-landau.de

A. Müller
University of Geneva, Geneva, Switzerland

R. Hettmannsperger
Goethe-University Frankfurt am Main, Frankfurt, Germany
e-mail: hettmannsperger@em.uni-frankfurt.de

© Springer International Publishing AG, part of Springer Nature 2018
K. L. Daniel (ed.), *Towards a Framework for Representational Competence in Science Education*, Models and Modeling in Science Education 11,
https://doi.org/10.1007/978-3-319-89945-9_13

used as a tool for problem solving tasks and is an essential means of reasoning. Schnotz and Mayer describe in detail the theory of cognitive processes that occur during the work with multiple representations (Schnotz 2005; Mayer 2005). So, information of different representations can be processed by the auditive or visual channel and integrated in propositions and mental models in the working memory (Schnotz 2005, Mayer 2005). Also, they provide task-construction strategies to reduce a possible cognitive overload (Sweller 1999) that can occur by the work with multiple representations (e.g., multimedia principle and coherence principle, Mayer 2005; Schnotz 2005).

The skilled use of several representations as problem-solving tools is well known in physics and science education (Ainsworth 1999; Tsui and Treagust 2013; Schnotz et al. 2011). The ability to generate and use different specific depictive or descriptive representations (Schnotz and Bannert 2003) of a situation or a problem in a skilled and interconnected way is called representational competence (RC, Guthrie 2002; Kozma and Russell 2005); the ability to change and translate between different forms of representations (Ainsworth 1999) and communicate about underlying, not obviously perceived, physical entities (e.g. radiation or atoms) and processes is included in the set of skills and practices of RC (Kozma and Russell 1997; Kozma 2000; Dolin 2007). A translation of information between representations is often necessary (Ainsworth 1999); in a scientific context, it means that some parts of the represented information can be expressed in different forms of representation. Experts perform significantly better than novices when translating the content of a graph, a video, or an animation about molecules into any other type of representation (Kozma and Russell 1997). Of course, the overlapping content of a set of representations should be translated without contradictions. Each representation type needs a specific way of thinking and has its assets and drawbacks; therefore a skilled combination of different representations is assumed to be beneficial for the learning process (Leisen 1998a; Leisen 1998b). In view of its importance for scientific thinking and based on a considerable body of evidence (e.g. Kozma et al. 2000; Dunbar 1997; Roth and McGinn 1998), Kozma (2000) emphasized that the development of RC should be included in the chemistry curriculum. In Denmark, it is already implemented in the school curriculum (Dolin 2007).

Many studies have shown that the levels of students RC are low. Kozma and Russell reported (2005) that students with different levels of RC show differences in their work with representations. They pointed out that persons with a low level of RC work on the surface level of a representation (Chi et al. 1981; diSessa et al. 1991; Kozma and Russell 1997), whereas those with a high level of RC show features of deep-level processing; for instance, they use a higher number of formal and informal representations to solve problems, make predictions, or explain phenomena (Dunbar 1997; Goodwin 1995; Kozma et al. 2000; Kozma and Russell 1997; Roth and McGinn 1998). Research has identified possible reasons for the low level of students' RC in chemistry (Devetak et al. 2004): Secondary school students (average age: 18 years) had to solve several tasks in their high school examination. To do this, they had to be able to connect macroscopic, submicroscopic, and symbolic levels of chemical concepts (Thiele and Treagust 1994; Devetak et al. 2004).

However, they concluded that teachers do not usually focus on teaching students how to connect several representations; they only concentrate on it during the preparation of the high school examination. If students are not able to connect levels of representations sufficiently, their knowledge is fragmented and can only be remembered temporarily. Additionally, it is known that the problem is not only a lack of students' ability to interconnect different representation levels; in addition, students do not clearly see the role of symbolic and submicroscopic levels of representation (Treagust et al. 2003). Research has shown that students of physics also have a low level of RC (Saniter 2003): Even advanced learners (seventh semester of university) were not able to connect the meanings of formulas with phenomena and with practical implementation in experiments better than less advanced learners in the fifth semester of university. Problems occurred especially when students tried to explain an experiment only with one type of representation (e.g., with the symmetric form of coulombs law for a single source point charge, Saniter 2003). Only when students made a connection with a representation on the phenomenological level, they were able to estimate the measurement value correctly. Other representational problems have also been found (Saniter 2003). When a student was not familiar with a topic of a representation level very well, he/she could not use it to solve the task, even if he/she had already worked on the content in the task directly before (Saniter 2003). A kind of content-specific blindness is presumed as a reason for this. A danger for students is that they continue to operate based on the surface-level features of representations (Chi et al. 1981; diSessa et al. 1991; Kozma and Russell 1997). Another possible reason for the low level of students' RC might be the way teachers deal with representations in classes. Lee (2009) analyzed lessons in three eighth-grade classes on ray optics and found an implicit, short, partly inaccurate and receptive way of using representations. Accordingly, the students are not sufficiently cognitively activated regarding representations (for an activating learning strategy for development of RC see Scheid et al. 2015; Scheid 2013).

Not only does the lack of RC lead to deficits in the learning process, the students have also not met the expectations of teachers with regard to learning with experiments (Novak 1990; Harlen 1999). In school lessons, students had only seldom the opportunity to speak about experiments or design or analyze them on their own. Therefore, they were not able to make connections to the experiments (Tesch 2005, seventh to ninth grade, video study about mechanics and electricity). For this reason, students' opportunities to process the different types of representations in greater depth were low. An appropriate level of understanding of science experiments generally requires a certain level of RC, because information is usually spread over several representations of different types or also the same type and has to be connected. In particular, this means the ability to connect the content of different representations with each other and to translate the overlapping contents from one representation into another is a key competence to achieve connected knowledge. This competence is called Representational Coherence Ability (RCA, Scheid et al. 2015; Scheid 2013). Translating information between several representations is inherently susceptible to misinterpretation or failure, which can lead to unwanted contradictions and inconsistencies. Therefore, a central part of RC can be seen in the above mentioned RCA as the level of students' ability to achieve consistency between the overlapping information of a set of representations, which is scientifically correct. RCA is essential for the use of multiple representations; it includes also translating of information between different types of representations or adapting representations, and has a fundamental connection to achievement in the subject matter (e.g., physics). The facts of the importance of representational abilities for the understanding and, simultaneously, the low representational abilities of the students (see above) lead to an urgent need of a diagnostic instrument for RCA. Therefore, the goal of the study is to develop test items for RCA and to probe it for reliability via psychometrical values.

Methods

The research took place in several grammar schools in Germany (federal state Rhineland-Palatinate). All schools were located in small towns. The topic of the study was physics, in particular the subject of ray optics and image forming by a convex lens. According to the curriculum, this topic is taught in the seventh grade. We had three measurement times: directly before and after six optics lessons, and six weeks after these lessons. The data were analyzed with classical methods of item analysis: factor analysis, α_C, item-total correlation, item difficulty and for the expert rating intra-class correlation. We recruited 488 students in 17 classes and six schools. The students were between 12 and 14 years of age ($M = 13.3$; $SD = 0.5$), 54% boys and 46% girls.

Design of the RCA Test

The ability to design and adequately work with representations is connected closely with physics achievement, because scientific representations are often domain specific and describe, depict, or relate to scientific content. The ability to handle representations is in the same time foundation to develop achievement of physics and a consequence of it. So they are interdependent and their development goes hand in hand. Therefore, RCA assessing tasks are inherently related to physics achievement, even though they focus on RCA. This focus is set by the multiplicity of experiment- or phenomenon-related representations. It was measured through lesson-related assessment tasks, which require working with two representations (item-type A) or measured through tasks that require working mostly with three or more representations (item-type B, both see Table 1). By combining these variants, the spectrum of RC from single or few representations to a system of interconnected representations can be covered. Items 1b, 1c, 4a and 7 are discussed in this chapter. For illustration of the different representations of the items, the test is added in the appendix.

There are six items of type A and nine of type B. The test has an open-answer format and is interval scaled (Scheid 2013; Scheid et al. 2017; complexity of the answers was considered, Kauertz 2008).

Table 1 Overview about different combinations of representation types of the RCA test items (*: information not urgently needed to solve the task): items 1b, 1c, 4a and 7 are discussed in the text, other items and its representations see appendix

Item	Item-type	No. of representations needed for solving the task	Text	Ray diagram	Formula	Table	Photograph / Realistic drawing
1a	A	2	1	1			
2			1			1	
3b			*	1	1		
4a			1	1			
7			1	1			
1b	B	≥3	2	2			
1c			2	2			
3a			1	1			
4b			2	1			
5a			2				2
5b			2				2
5ca			2				2
5cb			2				2
6			2	2	1		
8			2				2

Peter has conducted an image-forming experiment and has forgotten to note the focus of the lens.

The item size was 3.0 cm, the picture size 1.6 cm, the image distance 4.3 cm and the picture distance 2.3 cm.

Solve the task with a drawing.
What is the focus of the lens?

Convex Lens

Fig. 1 Example of an item of type A that assesses RCA with two representations (item 7): ray construction with given numeric parameters (see Fig. 3 for an example of a ray construction, Scheid 2013; Scheid et al. 2017)

As an example for item-type A, Fig. 1 shows a task that assesses RCA via performance with a representational focus and uses a text and a ray diagram (Table 1); students were asked to find out the focus of a convex lens by constructing a ray diagram of an image-formation process with given numeric parameters. Students were only able to solve the task if they understood the verbal part of the task, found out how to obtain the solution, and if they were able to translate the relevant part of the verbal information into a depictive schematic drawing and apply the physics rules of ray diagram construction. Figure 2 also shows an item that assesses RCA with two representations.

It looks like a traditional text task focusing on the magnification equation but solving the task also requires RCA (i.e., a level of coherence between the verbal representation and mathematical formula, Table 1). The students had to be able to allocate the verbally implemented numeric parameters to the correct physics quanti-

Ines would like to draw an enlarged image (20 mm) of a lady bug that originally has a size of 5 mm. To facilitate drawing, she wants to project an image of the bug onto a screen. The lady bug is located 10 mm in front of an appropriate lens.
At what distance should the screen be positioned?
Solve the task with a calculation.

Fig. 2 Example of an item of type A to assess RCA with two representations (item 4a): calculation with verbally implemented parameters (Scheid 2013; Scheid et al. 2017)

b) If the item-lens distance is increased, how will the picture-lens distance and the picture size change?

____________________________...

c) Explain your reasoning with the rays of the diagram: If the item-lense distance increases, why does the picture size change in the way that you described above?

____________________________...

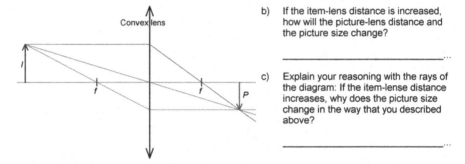

Fig. 3 Item-set for measurement of RCA with three or more representations (items 1b and 1c, Type B, Scheid 2013; Scheid et al. 2017; *I*: item size; *f*: focal length, *P*: picture size)

ties, otherwise they were not able to correctly insert the values into the formula and, therefore, not able to apply the formula correctly.

Also when students were asked to express the answer to the task verbally, they need to translate the mathematical representation into a text. For this purpose, they had to know at least the meaning of the physical parameters in the equation.

In contrast, items of type B measure RCA consisting of a set of tasks that normally require three or more coherent mental representations related to experiments or phenomena and therefore deep, under-the-surface processing to be able to solve the tasks.

Figure 3 shows an item set that required the student to develop a representational mental model to produce the correct answer. This model has to contain relevant information that is shown in the ray diagram of the experimental setting shown in the picture. To perform this, it is necessary to know the physics law concerning how certain rays can be changed by a convex lens if the object distance changes. This model was not usually available for students, but they had basic knowledge and therefore an opportunity to design the mental model. They know how to draw a ray diagram and the next step is to develop an appropriate mental model that compares and describes the outcomes of several ray diagrams with different image-lens distances and then to develop the required relation at the end of the representational thinking process (the ray construction had to alter mentally, Table 1). In this way it was possible to estimate how the image distance changes in connection with the object distance, both in particular and in general. In the second part of the task students were asked to explain verbally how they solved the tasks above. This approach attempted to gain an insight into their thinking processes and enabled the identification of students who randomly gave correct answers. For this reason, this task was only considered correct if a coherent mental model and three types of adequate representations were used: the schematic drawing of the printed task, a mental model of the schematic drawing in connection with a useful physics law (e.g., how the rays change), and a produced text describing the outcomes of the mental model as part of the answer.

Results of Item and Test Analysis

An exploratory factor analysis (principal component analysis, quartimax rotation, Eid et al. 2011) shows that the first factor has a large eigenvalue whereas the other factors have the level of individual items or are below that level. Factors can be seen as relevant, if they clearly show higher eigenvalues as the others (Cattell 1966) and the eigenvalues of the factors are higher than the ones of single items (Kaiser and Dickman 1959; Eid et al. 2011). So, only one factor is really relevant and the instrument can be seen as one dimensional.

With regard to reliability and validity for the curriculum, we obtained internal consistencies of the RCA test of $\alpha_C = 0.8$ (N(post) = 488, N(follow-up) = 484, Cronbach and Snow 1977). Pretest values cannot be expected to be in the desired ranges, as there is no consistent knowledge yet (Nersessian 1992; Ramlo 2008; Nieminen et al. 2010). Excluding individual items does not lead to an increase of the internal consistency. The item-total correlation of the RCA test was calculated, what means the correlation of single items with the total score of the remaining items. Every item has the desired correlation above $r_{it} > 0.3$ with the total score of the rest of the items (Weise 1975). Only item 3a has a correlation of $r_{it} = 0.2$.

In the pretest, the item difficulty is within the desired range of $0.2 < P_i < 0.8$ for eight items and lower than 0.2 for seven items ($\bar{P}_i(pretest) = 0.16$). For the other measurement times, no item has a difficulty outside the desired range except for item 1c (post and follow-up) and item 6 (follow-up). The mean item difficulty of the posttest (P_i (post) = 0.36) is 17% higher than the mean of the follow-up test ($\bar{P}_i$ (follow-up) = 0.30).

The results of the expert rating showed that the RCA test is seen as "valid for the curriculum" or "completely valid for the curriculum". The intra-rater correlations were highly significant and the values were between $0.5 < ICC < 0.7$.

Discussion

Regarding the item analysis, a possible explanation for the low item difficulties of the RCA pretest is that the physics content that was asked was actually the subject of the following lessons. Nevertheless, it made sense to assess RCA before the lessons started, because students showed variance in RCA in the pretest and this information could be used in the statistical analysis for measuring changes. The low item difficulties in the pretest and several missing values in the datasheet were the reason why the corrected item-total correlation and α_C could not be calculated for that time. Altogether three items were conspicuous at other measuring times. The item difficulty of item 1c is low because it required to logically reason about an abstract interrelation between the item distance and the image distance of an image formation experiment. Item 6 showed also a low item difficulty (but only in follow-up test); it asks for reasons why it is not possible to derive the magnification equation with two triangles that were marked in a ray diagram. For both items, the contents are difficult from the physics point of view and known to be difficult from the teaching practice; however, just because of these contents the items are interesting and important for measuring high levels of RCA. Therefore they are useful and may remain in the test. For item-total correlation, only item 3a had a low value and appeared to be an exception. A possible explanation could be that the topic of the item differed from the topics of the remaining items. It asked for the application possibilities of the magnification equation, therefore requiring metacognition. A remaining of the item in the instrument is questionable and it may be excluded. The values of the corrected item-total correlations and the value of α_C of the post- and follow-up tests were acceptable (Kline 2000). In sum, the RCA test can be considered as reliable.

The intra-class correlation of the expert rating was highly significant with acceptable values (Wirtz and Caspar 2002). So the expert rating showed clearly that the RCA test is valid for the curriculum and can, therefore, be used in schools to diagnose students' individual levels of RCA and also, if needed, for grading purposes.

Implications, Limitations, and Recommendations for Future Research

The development of a theory-based strategy for designing items for a RCA instrument was successful in respect to the considered aspects. So, with knowledge about the outcomes of this study, instructors can either generally assess whether teaching lessons fostering RCA are necessary or, in particular, identify which students need individual help to develop RCA and how much help they need. This can help to overcome well-known problems students face in understanding science concepts, phenomena, and experiments. The test instrument for RCA allows also to investigate the effects of future strategies to foster the development of RCA (e.g. by using RATs, see Scheid et al. 2015; Scheid 2013). A limitation is that the instrument is only available for the domain of ray optics. We recommend the design of RCA tests for other topics of science education that use multiple representations, as example for thermodynamics or genetics.

Acknowledgments We thank all teachers and students from the schools for participating in the study and the federal state Rhineland-Palatinate for the permission to realize it. We are grateful to the German Research Association (DFG, Graduate School GK1561), who funded this research. The opinions reported do not represent the views of the funding body.

Appendix

Items and representations of the RCA instrument

Task 1

a) Draw the propagation of the rays in the following image formation experiment:

$f = 3$ cm   item width $i = 8$ cm   item size $I = 2$ cm

b) If the item-lens distance is increased, how will the picture-lens distance and the picture size change?

_____
_____
_____

c) Explain your reasoning with the rays of the diagram: If the item-lens distance increases, why does the picture size change in the way that you described above?

_____
_____
_____
_____

Task 2

An item is in front of a convex lens. The item width is between the simple focal width and the double focal width.

What size has the picture compared to the item size? _____

At what distance to the lens (picture width) will the picture appear compared to the focal width? _____

Task 3

a) For what purposes is the magnification equation useful?

- To be able to design a lens experiment with a certain magnification factor, if picture width and picture size are known.

- To calculate the picture width, if item size and picture size are known.

- To calculate the magnification ratio without conducting a real experiment, if item width and picture width are known.

- To calculate the ratio of picture size and item size, if the focal width and image width are known.

b) Calculate the missing values of a projection of an item onto a screen. Write down the calculation in detail with its units.

I Item size	i Item width	P Picture size	p Picture width	A Projection scale
12 cm	120 cm		40 cm	

Task 4

Ines would like to draw an enlarged image (20 mm) of a lady bug that originally has a size of 5 mm. To facilitate drawing, she wants to project an image of the bug onto a screen. The lady bug is located 10 mm in front of an appropriate lens.
At what distance should the screen be positioned?
a. Solve the task with a calculation.

b. If the focal width of the lens was known, what possibility to solve the task would be available? Write only catchwords.

Task 5

A common window is projected by a convex lens to the wall of a room.

a) Which type of image (formed by a convex lens) is shown in the photograph below:
    enlarged

b) /reduced/equal image size?

_____

c) How large is the distance i between the object (real window) and the convex lens? Tick the
    right answer below.

         o i is approximately so far like the distance between wall and lens.
         o i is nearer than the distance between wall and lens
         o i is much farther than the distance between wall and lens.
         o i has double the distance between wall and lens

d) Explain your reasoning for question a) and b) in a few sentences.
       Question a) _____

       _____
       _____

       Question b):_____

       _____
       _____
       _____

Task 6

Why it is not possible to set up the projection equation with the following triangles?

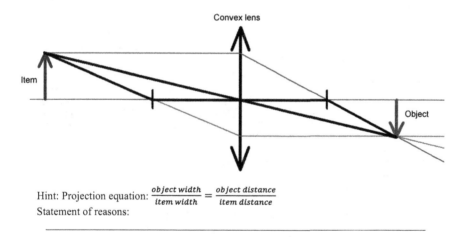

Convex lens

Item

Object

Hint: Projection equation: $\dfrac{object\ width}{item\ width} = \dfrac{object\ distance}{item\ distance}$

Statement of reasons:

_____
_____
_____
_____

Task 7

Peter has conducted an image-forming experiment and has forgotten to note the focus of
the lens. The item size was 3.0 cm, the picture size 1.6 cm, the image distance 4.3 cm and
the picture distance 2.3 cm.
Solve the task with a drawing. What is the focus of the lens?

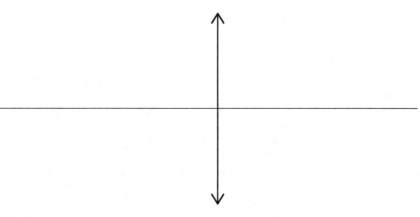

Task 8

On a sunny day, Tobias would like to inflame a match by a convex lens.

Lens   Match

Nadine proposed Tobias to place an adjustable diaphragm (see figure below) in front of the lens.

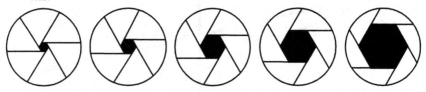

Is Tobias able to inflame the match easier by that?

No, because...

_____

If no, what setting of the diaphragm limits least? Mark in the figure above.

Yes, because...

_____

If yes, what setting of the diaphragm improves it best? Mark in in the figure above.

(all items of the RCA instrument translated from Scheid, 2013)

References

Ainsworth, S. (1999). The functions of multiple representations. *Computers & Education, 33,* 131–152.

Cattell, R. B. (1966). The scree test for the number of factors. *Multivariate Behavioral Research, 1,* 245–276.

Chi, M., Feltovich, P., & Glaser, R. (1981). Categorization and representation of physics problems by experts and novices. *Cognitive Science, 5,* 121–152.

Cronbach, L., & Snow, R. (1977). *Aptitudes and instructional methods: A handbook for research on interactions.* New York: Irvington.

Devetak, I., Urbančič, M. W., Grm, K. S., Krnel, D., & Glažar, S. A. (2004). Submicroscopic representations as a tool for evaluating students' chemical conceptions. *Acta Chimica Slovenica, 51,* 799–814.

Dolin, J. (2007). Science education standards and science assessment in Denmark. In D. Waddington, P. Nentwig, & S. Schanze (Eds.), *Making it comparable. Standards in science education* (pp. 71–82). Waxmann: Münster.

Dunbar, K. (1997). How scientists really reason: Scientific reasoning in real-world laboratories. In R. Sternberg & Davidson (Eds.), *The nature of insight* (pp. 365–396). Cambridge, MA: MIT Press.

Eid, M., Gollwitzer, M., & Schmitt, M. (2011). *Statistik und Forschungsmethoden. (2. Aufl.) [Statistics and research methods]* (2nd ed.). Weinheim: Beltz.

Gilbert, J. K., & Treagust, D. (Eds.). (2009). *Multiple representations in chemical education.* The Netherlands: Springer.

Goodwin, C. (1995). Seeing in depth. *Social Studies of Science, 25,* 237–274.

Guthrie, J. W. (Ed.). (2002). *Encyclopedia of education.* New York: Macmillan.

Harlen, W. (1999). *Effective teaching of science.* Edinburgh: The Scottish Council for Research in Education (SCRE).

Herrmann, T. (1993). Mentale Repräsentation ein erläuterungsbedürftiger Begriff [mental representation, an explanation needy expression]. In J. Engelkamp & T. Pechmann (Eds.), *Mentale Repräsentation* (pp. 17–30). Huber: Bern.

Kaiser, H. F., & Dickman, K. W. (1959). Analytic determination of common factors. *American Psychologist, 14,* 425.

Kauertz, A. (2008). Schwierigkeitserzeugende Merkmale physikalischer Leistungstestaufgaben [Difficulty-generating features of physics assessment tasks]. In H. Niedderer, H. Fischler, & E. Sumfleth (Eds.), *Studien zum Physik- und Chemielernen Band 79.* Berlin: Logos Verlag.

Kline, P. (2000). *The handbook of psychological testing* (2nd ed.). London: Routledge.

Kozma, R. (2000). Representation and language: The case for representational competence in the chemistry curriculum. *Paper presented at the Biennial Conference on Chemical education,* Ann Arbor, MI.

Kozma, R. B., & Russell, J. (1997). Multimedia and understanding: Expert and novice responses to different representations of chemical phenomena. *Journal of Research in ScienceTeaching, 43*(9), 949–968.

Kozma, R. B., & Russell, J. (2005). Students becoming chemists: Developing representational competence. In J. Gilbert (Ed.), *Visualization in science education.* Dordrecht: Springer.

Kozma, R. B., Chin, E., Russell, J., & Marx, N. (2000). The role of representations and tools in the chemistry laboratory and their implications for chemistry learning. *Journal of the Learning Sciences, 9*(2), 105–143.

Lee, V. (2009). Examining patterns of visual representation use in middle school science classrooms. *Proceedings of the National Association of research in science teaching (NARST) annual meeting compact disc*, Garden Grove, CA: Omnipress.

Leisen, J. (1998a). Physikalische Begriffe und Sachverhalte. Repräsentationen auf verschiedenen Ebenen [physical notions and concepts. Representations on different levels]. *Praxis der Naturwissenschaften Physik, 47*(2), 14–18.

Leisen, J. (1998b). Förderung des Sprachlernens durch den Wechsel von Symbolisierungsformen im Physikunterricht [Fostering the learning of conversation via the change of symbolizitation-types]. *Praxis der Naturwissenschaften Physik, 47*(2), 9–13.

Mayer, R. E. (2005). Cognitive theory of multimedia learning. In R. E. Mayer (Ed.), *The Cambridge handbook of multimedia learning* (pp. 31–48). New York: Cambridge University Press.

Nersessian, N. J. (1992). How do scientists think? Capturing the dynamics of conceptual change in science. In R. N. Giere (Ed.), *Cognitive models of science, Minnesota studies in the philosophy of science* (Vol. 15, pp. 129–186). Minneapolis: University of Minnesota Press.

Nieminen, P., Savinainen, A., & Viiri, J. (2010). Force concept inventory-based multiple-choice test for investigating students' representational consistency. *Phyical Review Special Topics Physics Education Research, 6*(2), 1–12.

Novak, J. D. (1990). The interplay of theory and methodology. In E. Hegarty-Hazel (Ed.), *The student laboratory and the science curriculum*. London, New York: Routledge.

Ramlo, S. (2008). Validity and reliability of the force and motion conceptual evaluation. *American Journal of Physics, 76*(9), 882–886.

Roth, W. M., & McGinn, M. (1998). Inscriptions: A social practice approach to representations. *Review of Educational Research, 68*, 35–59.

Saniter, A. (2003). Spezifika der Verhaltensmuster fortgeschrittener Studierender der Physik [The specifics of behavior patterns in advanced students of physics]. In H. Niedderer & H. Fischler (Eds.), *Studien zum Physiklernen Band 28*. Berlin: Logos.

Scheid, J. (2013). Multiple Repräsentationen, Verständnis physikalischer Experimente und kognitive Aktivierung: Ein Beitrag zur Entwicklung der Aufgabenkultur [Multiple representations, understanding physics experiments and cognitive activation: A contribution to developing a task culture]. In H. Niedderer, H. Fischler, & E. Sumfleth (Eds.), *Studien zum Physik- und Chemielernen, Band 151*. Logos Verlag: Berlin, Germany.

Scheid, J., Müller, A., Hettmannsperger, R., & Schnotz, W. (2015). Scientific experiments, multiple representations, and their coherence. A task based elaboration strategy for ray optics. In W. Schnotz, A. Kauertz, H. Ludwig, A. Müller, & J. Pretsch (Eds.), *Multiple perspectives on teaching and learning*. Basingstoke: Palgrave Macmillan.

Scheid, J., Müller A., Hettmannsperger, R. & Kuhn, J. (2017). Erhebung von repräsentationaler Kohärenzfähigkeit von Schülerinnen und Schülern im Themenbereich Strahlenoptik [Assessment of Students Representational Coherence Ability in the Area of Ray Optics]. *Zeitschrift für Didaktik der Naturwissenschaften, 23*, 181–203.

Schnotz, W. (1994). *Aufbau von Wissensstrukturen. Untersuchungen zur Kohärenzbildung bei Wissenserwerb mit Texten [Construction of knowledge structures. Research on coherence development during knowledge acquisition with texts]*. Beltz, Psychologie-Verl.Union: Weinheim.

Schnotz, W. (2002). Towards an integrated view of learning from text and visual displays. *Educational Psychology Review, 14*(1), 101–119.

Schnotz, W. (2005). An integrated model of text and picture comprehension. In R. E. Mayer (Ed.), *The Cambridge handbook of multimedia learning* (pp. 49–70). New York: Cambridge University Press.

Schnotz, W., & Bannert, M. (2003). Construction and interference in learning from multiple representation. *Learning and Instruction, 13*(2), 141–156.

Schnotz, W., Baadte, C., Müller, A., & Rasch, R. (2011). Kreatives Denken und Problemlösen mit bildlichen und beschreibenden Repräsentationen [Creative thinking and problem solving with depictoral and descriptive representations]. In R. Sachs-Hombach & R. Totzke (Eds.), *Bilder Sehen Denken* (pp. 204–252). Köln: Halem Verlag.

diSessa, A., Hammer, D., Sherin, B., & Kolpakowski, T. (1991). Inventing graphing: Metarepresentational expertise in children. *Journal of Mathematical Behavior, 10*(2), 7–160.

Sweller, J. (1999). *Instructional design in technical areas.* Melbourne: ACER Press.

Tesch, M. (2005). Das experiment im Physikunterricht [The experiment in physics education]. In H. Niedderer, H. Fischler, & E. Sumfleht (Eds.), *Studien zum Physik- und Chemielernen Band 42.* Berlin: Logos.

Thiele, R. B., & Treagust, D. F. (1994). An interpretive examination of high school chemistry teachers' analogical explanations. *Journal of Research in Science Teaching, 31,* 227–242.

Treagust, D. F., Chittleborough, G. D., & Mamiala, T. L. (2003). The role of sub-microscopic and symbolic representations in chemical explanations. *International Journal of Science Education, 25*(11), 1353–1369.

Tsui, C., & Treagust, D. (Eds.). (2013). *Multiple representations in biological education.* Dordrecht: Springer.

Tytler, R., Prain, V., Hubber, P., & Waldrip, B. (Eds.). (2013). *Constructing representations to learn in science.* Rotterdam: Sense.

Weise, G. (1975). *Psychologische Leistungstests [Psychological performance tests].* Göttingen: Hogrefe.

Wirtz, M., & Caspar, F. (2002). *Beurteilerübereinstimmung und Beurteilerreliabilität [Rater agreement and rater reliability].* Göttingen: Hogrefe.

Printed in the United States
By Bookmasters